Collins

The Shanghai Maths Project

For the English National

一课一练

Practice Book 3B

Series Editor: Professor Lianghuo Fan

UK Curriculum Consultant: Paul Broadbent

T0173427

William Collins' dream of knowledge for all began with the publication of his first book in 1819.

A self-educated mill worker, he not only enriched millions of lives, but also founded a flourishing publishing house. Today, staying true to this spirit, Collins books are packed with inspiration, innovation and practical expertise. They place you at the centre of a world of possibility and give you exactly what you need to explore it.

Collins. Freedom to teach.

Published by Collins
An imprint of HarperCollins*Publishers*
The News Building
1 London Bridge Street
London
SE1 9GF

HarperCollins *Publishers*
1st Floor
Watermarque Building
Ringsend Road
Dublin 4
Ireland

Browse the complete Collins catalogue at
www.collins.co.uk

Series Editor: Professor Lianghuo Fan
UK Curriculum Consultant: Paul Broadbent
Publishing Manager: Fiona McGlade
In-house Editor: Nina Smith
In-house Editorial Assistant: August Stevens
Project Manager: Emily Hooton
Copy Editors: Catherine Dakin and Tanya Solomons
Proofreader: Helen Pettitt
Cover design: Kevin Robbins and East China Normal University Press Ltd.
Cover artwork: Daniela Geremia

MIX
Paper from
responsible sources
FSC www.fsc.org **FSC™ C007454**

This book is produced from independently certified FSC paper to ensure responsible forest management.

For more information visit:
www.harpercollins.co.uk/green

Internal design: 2Hoots Publishing Services Ltd
Typesetting: 2Hoots Publishing Services Ltd
Illustrations: QBS
Production: Rachel Weaver
Printed and Bound in the UK using 100% Renewable Electricity at CPI Group (UK) Ltd

The Shanghai Maths Project (for the English National Curriculum) is a collaborative effort between HarperCollins, East China Normal University Press Ltd. and Professor Lianghuo Fan and his team. Based on the latest edition of the award-winning series of learning resource books, *One Lesson, One Exercise*, by East China Normal University Press Ltd. in Chinese, the series of Practice Books is published by HarperCollins after adaptation following the English National Curriculum.

Practice Book Year 3B has been translated and developed by Professor Lianghuo Fan with the assistance of Ellen Chen, Ming Ni, Huiping Xu and Dr Lionel Pereira-Mendoza, with Paul Broadbent as UK Curriculum Consultant.

Contents

Chapter 7 Addition and subtraction with 3-digit numbers

7.1   Addition and subtraction of whole hundreds and tens (1) . . 1

7.2   Addition and subtraction of whole hundreds and tens (2) . . 4

7.3   Adding and subtracting 3-digit numbers and ones (1) 8

7.4   Adding and subtracting 3-digit numbers and ones (2) 11

7.5   Addition with 3-digit numbers (1) 14

7.6   Addition with 3-digit numbers (2) 17

7.7   Subtraction with 3-digit numbers (1) 20

7.8   Subtraction with 3-digit numbers (2) 23

7.9   Estimating addition and subtraction with 3-digit numbers (1) 26

7.10   Estimating addition and subtraction with 3-digit numbers (2) 29

Chapter 7 test 33

Chapter 8 Simple fractions and their addition and subtraction

8.1   Unit fractions and tenths 38

8.2   Non-unit fractions 42

8.3   Equivalent fractions 45

8.4   Addition and subtraction of simple fractions 48

Chapter 8 test 52

Chapter 9 Multiplying and dividing by a 1-digit number

9.1   Multiplying by whole tens and hundreds (1) 56

9.2   Multiplying by whole tens and hundreds (2) 59

9.3   Writing number sentences 62

9.4   Multiplying a 2-digit number by a 1-digit number (1) 65

9.5   Multiplying a 2-digit number by a 1-digit number (2) 68

9.6   Multiplying a 2-digit number by a 1-digit number (3) 71

9.7   Multiplying a 3-digit number by a 1-digit number (1) 74

9.8   Multiplying a 3-digit number by a 1-digit number (2) 77

9.9   Practice and exercise 80

9.10   Dividing whole tens and whole hundreds 83

9.11   Dividing a 2-digit number by a 1-digit number (1). 87

9.12   Dividing a 2-digit number by a 1-digit number (2). 90

9.13   Dividing a 2-digit number by a 1-digit number (3). 93

9.14   Dividing a 2-digit number by a 1-digit number (4). 97

9.15   Dividing a 2-digit number by a 1-digit number (5). 101

9.16   Dividing a 3-digit number by a 1-digit number (1). 105

9.17   Dividing a 3-digit number by a 1-digit number (2). 108

9.18   Dividing a 3-digit number by a 1-digit number (3). 112

9.19   Application of division 115

9.20   Finding the total price 118

Chapter 9 test. 121

Chapter 10 Let's practise geometry

10.1 Angles. 126

10.2 Identifying different types of line (1) 129

10.3 Identifying different types of line (2) 132

10.4 Drawing 2-D shapes and making 3-D shapes 135

10.5 Length: metre, centimetre and millimetre 139

10.6 Perimeters of simple 2-D shapes (1). 143

10.7 Perimeters of simple 2-D shapes (2). 146

Chapter 10 test 150

End of year test 155

Chapter 7 Addition and subtraction with 3-digit numbers

7.1 Addition and subtraction of whole hundreds and tens (1)

Learning objective Add and subtract multiples of 10 and 100

Basic questions

1 Fill in the boxes.

(a) 300 + 200 means ☐ hundreds plus ☐ hundreds, which makes ☐ hundreds.

(b) 400 + 500 means ☐ hundreds plus ☐ hundreds, which makes ☐ hundreds.

(c) 900 − 300 means ☐ hundreds minus ☐ hundreds, which makes ☐ hundreds.

(d) 600 − 400 means ☐ hundreds minus ☐ hundreds, which makes ☐ hundreds.

2 Calculate mentally.

(a) 500 + 200 = ☐　　　　(b) 900 − 300 = ☐

(c) 400 − 100 = ☐　　　　(d) 200 + 400 = ☐

(e) 300 + 600 = ☐　　　　(f) 300 + 200 = ☐

(g) 400 − 300 = ☐　　　　(h) 500 − 100 = ☐

3 Fill in the boxes.

(a) 450 + 20 means ☐ tens plus ☐ tens, which makes ☐ tens.

(b) 450 − 20 means ☐ tens minus ☐ tens, which makes ☐ tens.

(c) 360 + 120 means ☐ tens plus ☐ tens, which makes ☐ tens.

(d) 360 − 120 means ☐ tens minus ☐ tens, which makes ☐ tens.

4 Calculate mentally.

(a) 12 + 7 = ☐   (b) 120 + 70 = ☐

(c) 45 − 8 = ☐   (d) 450 − 80 = ☐

(e) 35 + 8 = ☐   (f) 350 + 80 = ☐

(g) 91 − 5 = ☐   (h) 910 − 50 = ☐

5 Write >, < or = in each ◯.

(a) 800 + 300 ◯ 300 + 800   (b) 460 − 40 ◯ 460 + 40

(c) 540 + 70 ◯ 40 + 570   (d) 690 − 90 ◯ 700 − 90

6 Application problems.

(a) A washing machine costs £230. A TV costs £400. How much cheaper is the washing machine than the TV?

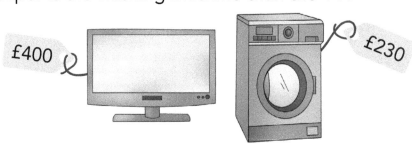

£400

£230

Answer: _____

(b) There are 530 story books and 380 science books in a library. How many story books and science books are there in total?

Answer: _____

(c) Finn's father bought him a bike for £180. He paid the shop assistant £200. How much change should he have received?

Answer: _____

(d) There are 340 cherry trees in an orchard, which is 270 fewer than the number of apple trees. How many apple trees are there?

Answer: _____

Challenge and extension question

7 Which is greater, ● or ■? How much greater is it?

$$● + 150 = ■ − 150$$

7.2 Addition and subtraction of whole hundreds and tens (2)

Learning objective Add and subtract multiples of 10 and 100

Basic questions

1 Calculate mentally.

(a) 870 − 700 = ☐

(b) 500 + 320 = ☐

(c) 760 − 560 = ☐

(d) 900 − 190 = ☐

(e) 730 + 150 = ☐

(f) 670 + 300 = ☐

(g) 370 − 150 = ☐

(h) 640 + 90 = ☐

(i) 130 + 370 = ☐

(j) 360 − 190 = ☐

(k) 480 − 250 = ☐

(l) 1000 − 650 = ☐

2 Complete the tables.

(a)

Addend	280	210	330	140	240	390	190
Addend	230	160	150	360	410	220	810
Sum							

(b)

Minuend	160	240	430	650	710	480	610
Subtrahend	80	90	210	360	450	290	450
Difference							

3 Write the sum and the difference of the two numbers on each card.

(a)

600
200   Sum: ☐

Difference: ☐

(b)

250
80   Sum: ☐

Difference: ☐

(c)

460
170   Sum: ☐

Difference: ☐

(d)

720
280   Sum: ☐

Difference: ☐

4 Calculate and then fill in each box with your answer.

(a) $120 \xrightarrow{+\ 90}$ ☐ $\xrightarrow{+\ 270}$ ☐ $\xrightarrow{+\ 160}$ ☐ $\xrightarrow{+\ 190}$ ☐

(b) $870 \xrightarrow{-\ 70}$ ☐ $\xrightarrow{-\ 350}$ ☐ $\xrightarrow{+\ 320}$ ☐ $\xrightarrow{-\ 270}$ ☐

(c) $330 \xrightarrow{+\ 120}$ ☐ $\xrightarrow{-\ 80}$ ☐ $\xrightarrow{+\ 290}$ ☐ $\xrightarrow{-\ 340}$ ☐

(d) ☐ $\xrightarrow{+\ 360}$ ☐ $\xrightarrow{-\ 150}$ ☐ $\xrightarrow{+\ 110}$ ☐ $\xrightarrow{-\ 250}$ 100

5 Application problems.

(a) Meera and Caitlin took part in an 800 metre run. When Meera was 200 metres away from the finish line, Caitlin was 250 metres away. How many metres had each of them run? Who had run faster up to that point in time?

Answer: _____

(b) A school bought 250 apples and gave 90 of them to Year 3 and 110 to Year 4. How many apples were left?

Answer: _____

(c) Tom collected 150 stamps last year.
He has collected 280 stamps this year, but he still has 130 stamps fewer than Yee. How many stamps has Yee collected?

Answer: _____

(d) A small cinema has 30 seats. It has 80 seats fewer than a medium-sized cinema.
A large cinema has 160 seats more than the medium-sized cinema. How many seats does the large cinema have?

Answer: _____

6   Look at the diagram. Fill in each circle with these numbers so that the sum of the four numbers in the corners of each square is 1200.

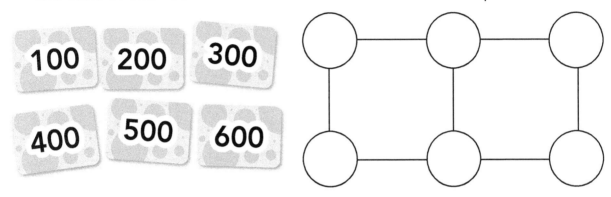

7.3 Adding and subtracting 3-digit numbers and ones (1)

 Learning objective Add and subtract ones from 3-digit numbers

 Basic questions

1 Calculate with reasoning.

(a) 23 + 2 =

(b) 46 − 8 =

(c) 52 − 5 =

(d) 26 + 8 =

(e) 323 + 2 =

(f) 246 − 8 =

(g) 552 − 5 =

(h) 426 + 8 =

2 Calculate mentally.

(a) 326 + 8 =

(b) 113 − 5 =

(c) 312 + 8 =

(d) 119 − 8 =

(e) 223 − 4 =

(f) 233 + 8 =

(g) 450 − 6 =

(h) 592 + 8 =

3 Complete these addition and subtraction calculations using the number lines.

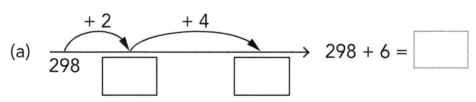

(a)   298 + 6 =

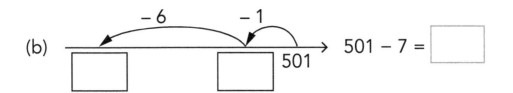

(b) 501 − 7 = ☐

−6   −1

☐   ☐ 501

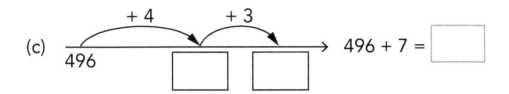

(c) 496 + 7 = ☐

+ 4   + 3

496   ☐   ☐

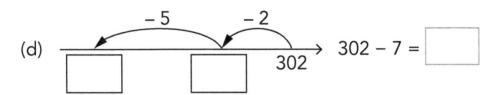

(d) 302 − 7 = ☐

− 5   − 2

☐   ☐ 302

4 Complete the tables.

(a)
+ 6

395	
597	

(b)
+ 8

793	
899	

(c)
− 6

201	
704	

(d)
− 8

603	
505	

5 Calculate mentally.

(a) 289 + 6 = ☐   (b) 203 − 5 = ☐   (c) 495 + 6 = ☐

(d) 910 − 7 = ☐   (e) 320 − 4 = ☐   (f) 698 + 4 = ☐

(g) 790 − 6 = ☐   (h) 589 + 2 = ☐   (i) 503 − 4 = ☐

(j) 763 + 8 = ☐   (k) 494 + 6 = ☐   (l) 388 + 3 = ☐

6 Calculate and then fill in each box with your answer.

(a) 228 $\xrightarrow{+6}$ ☐ $\xrightarrow{+7}$ ☐ $\xrightarrow{+8}$ ☐ $\xrightarrow{+9}$ ☐

(b) 397 $\xrightarrow{+8}$ ☐ $\xrightarrow{-5}$ ☐ $\xrightarrow{+3}$ ☐ $\xrightarrow{-6}$ ☐

(c) 689 $\xrightarrow{+6}$ ☐ $\xrightarrow{+8}$ ☐ $\xrightarrow{-9}$ ☐ $\xrightarrow{-4}$ ☐

 Challenge and extension question

7 Use the numbers shown on the cards below to form number sentences and then calculate.

5   3   7   9   0

(a) Make two addition sentences to add a 3-digit number and a 1-digit number.

(i) ☐ + ☐ = ☐

(ii) ☐ + ☐ = ☐

(b) Make two subtraction sentences to subtract a 1-digit number from a 3-digit number.

(i) ☐ + ☐ = ☐

(ii) ☐ + ☐ = ☐

7.4 Adding and subtracting 3-digit numbers and ones (2)

Learning objective Add and subtract ones from 3-digit numbers

Basic questions

1 Calculate mentally.

(a) 325 + 8 = ☐   (b) 590 − 9 = ☐   (c) 708 + 9 = ☐

(d) 377 − 8 = ☐   (e) 278 − 6 = ☐   (f) 498 + 6 = ☐

(g) 300 − 3 = ☐   (h) 799 + 2 = ☐   (i) 702 − 4 = ☐

(j) 375 + 5 = ☐   (k) 256 + 6 = ☐   (l) 173 + 7 = ☐

2 Fill in the blanks.

(a)

239 ☐

245

(b)

798 ☐

803

(c)

☐ 656

660

(d)

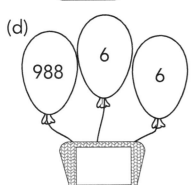

988   6   6

(e)

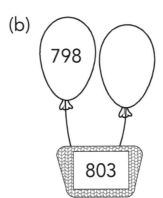

597   4   5

(f)

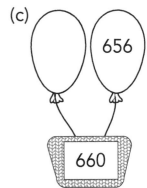

658   7

672

3 Draw lines to match the calculations with the correct answers.

(a) | 627 + 5 | | 702 | | 702 – 7 |

(b) | 689 + 6 | | 632 | | 710 – 8 |

(c) | 697 + 5 | | 695 | | 641 – 9 |

4 Application problems.

(a) During an environment project some Year 2 pupils collected 203 plastic containers, which was 9 fewer than the number collected by Year 3. How many containers did the Year 3 pupils collect?

Answer: _____

(b) Dillon is 132 cm tall. He is 5 cm taller than Asha. How tall is Asha?

Answer: _____

(c) An appliance shop had 256 TVs. They sold 8 TVs in the morning and 7 in the afternoon. How many TVs did the shop have left?

Answer: _____

(d) Max, May and Jo were skipping. Max did 142 skips, which was 6 skips fewer than Jo. May did 8 skips fewer than Jo. How many skips did May do?

Answer: _____

 Challenge and extension question

5 Look at the digit cards and answer the questions.

$$4 \quad 8 \quad 9 \quad 2$$

(a) What is the greatest number that can be made from the sum of a 3-digit number and a 1-digit number made using these digit cards? Write the number sentence.

_____

(b) What is the smallest number that can be made from the difference of a 3-digit number and a 1-digit number made using these digit cards? Write the number sentence.

_____

7.5 Addition with 3-digit numbers (1)

Learning objective Use partitioning to add 3-digit numbers

Basic questions

1 Calculate with reasoning.

(a) Theo's method.

322 + 216 = ☐

Hundreds + Hundreds:	300	+	200	=	☐
Tens + Tens:	20	+	10	=	☐
Ones + Ones:	2	+	6	=	☐
	☐	+	☐	+ ☐ =	☐

(b) Maya's method.

132 + 454 = ☐

Ones + Ones:	☐	+	☐	=	☐
Tens + Tens:	☐	+	☐	=	☐
Hundreds + Hundreds:	☐	+	☐	=	☐
	☐	+	☐	+ ☐ =	☐

(c) Suraj's method.

$$124 + 259$$

$$= 124 + 200 + 50 + 9$$

$$= \boxed{} + \boxed{} + \boxed{}$$

$$= \boxed{} + \boxed{}$$

$$= \boxed{}$$

$$379 + 146$$

$$= 379 + \boxed{} + \boxed{} + \boxed{}$$

$$= \boxed{} + \boxed{} + \boxed{}$$

$$= \boxed{} + \boxed{}$$

$$= \boxed{}$$

(d) Minna's method.

$$430 + 352$$

$$= 430 + 2 + 50 + 300$$

$$= \boxed{} + \boxed{} + \boxed{}$$

$$= \boxed{} + \boxed{}$$

$$= \boxed{}$$

$$182 + 219$$

$$= 182 + \boxed{} + \boxed{} + \boxed{}$$

$$= \boxed{} + \boxed{} + \boxed{}$$

$$= \boxed{} + \boxed{}$$

$$= \boxed{}$$

2 Use your preferred method to calculate. Show your working.

(a)
$$536 + 121$$

(b)
$$428 + 236$$

(c)
$$650 + 328$$

(d)
$$418 + 365$$

3 Use the six digit cards to form addition sentences adding two 3-digit numbers. Use your preferred method to work them out.

6 0 7 3 5 2

(a)

(b)

(c)

4 Write a number sentence for each question.

(a) One addend is 239 and the other is 384. What is the sum?

Number sentence: _____

(b) What number is 168 more than 574?

Number sentence: _____

Challenge and extension question

5 Fill in the boxes.

(a) The sum of the smallest 3-digit number and the smallest 3-digit number is a ☐-digit number.

(b) The sum of the greatest 3-digit number and the greatest 3-digit number is a ☐-digit number.

(c) The sum of two 3-digit numbers can be a ☐-digit number or a ☐-digit number.

7.6 Addition with 3-digit numbers (2)

Learning objective Use the column method to add 3-digit numbers

 Basic questions

1 Use the column method to calculate.

(a) 450 + 234 =

```
    4  5  0
 +  2  3  4
 _____

 _____
```

(b) 308 + 126 =

```
    3  0  8
 +  1  2  6
 _____

 _____
```

(c) 703 + 224 =

```
    7  0  3
 +  2  2  4
 _____

 _____
```

(d) 372 + 143 =

(e) 346 + 251 =

(f) 597 + 188 =

2 Are these calculations correct? Put a ✓ for yes or a ✗ for no in the box and then make corrections.

(a)
```
    5  8  1
 +  3  3
 _____
    9  1  1
 _____
```
☐

(b)
```
    3  7  8
 +  1  2  6
 _____
    4  9  4
 _____
```
☐

(c)
```
    3  5  4
 +  1  2  8
 _____
    4  8  2
 _____
```
☐

Corrections:

3 Complete the table.

Addend	327	204	534	178	257	689
Addend	150	328	265	433	465	311
Sum						

4 Fill in the boxes.

(a)
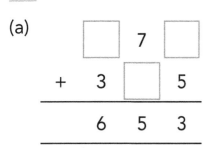

(b)
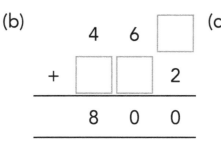

(c)

5 Application problems.

(a) An electrician cut 312 metres from a coil of electric wire. He then cut off another 268 metres. How many metres did he cut off in total?

Ans wer: _____

(b) Ollie was reading a book. In the first week he read 162 pages. This was 135 fewer pages than in the second week. How many pages did he read in the second week?

Answer: _____

(c) At Laptop World, a used laptop computer is priced at £138. A new laptop is priced at £162 more. What is the price of the new laptop?

 £138

Answer: _____

 Challenge and extension question

6 When Alisa was doing an addition sentence, by mistake she thought the digit 3 in the hundreds place was 2 and the digit 7 in the ones place was 1. The sum she got was 675. What is the correct sum?

The correct sum is: _____.

7.7 Subtraction with 3-digit numbers (1)

 Learning objective Use partitioning to subtract 3-digit numbers

 Basic questions

1 Calculate with reasoning.

(a) Theo's method.

467 − 253 = ☐

Hundreds – Hundreds:	400	– 200	= ☐
Tens – Tens:	60	– 50	= ☐
Ones – Ones:	7	– 3	= ☐
	☐ – ☐ – ☐	= ☐	

(b) Maya's method.

856 − 543 = ☐

Subtract hundreds first:	856	– 500	= ☐
Then subtract tens:	☐	– ☐	= ☐
Finally subtract ones:	☐	– ☐	= ☐
	☐ + ☐ + ☐	= ☐	

(c) Suraj's method.

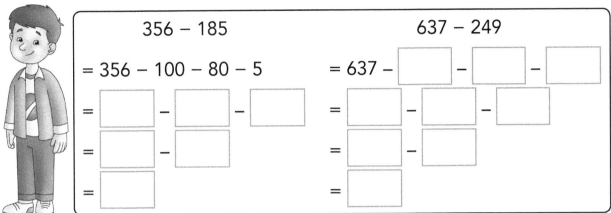

356 − 185

= 356 − 100 − 80 − 5

= ☐ − ☐ − ☐

= ☐ − ☐

= ☐

637 − 249

= 637 − ☐ − ☐ − ☐

= ☐ − ☐ − ☐

= ☐ − ☐

= ☐

(d) Minna's method.

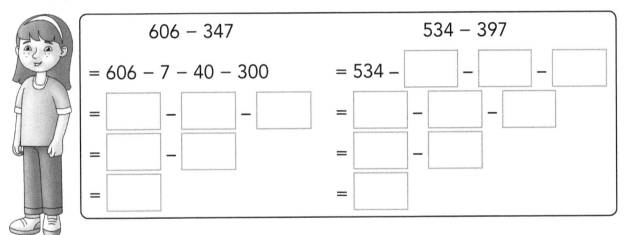

606 − 347

= 606 − 7 − 40 − 300

= ☐ − ☐ − ☐

= ☐ − ☐

= ☐

534 − 397

= 534 − ☐ − ☐ − ☐

= ☐ − ☐ − ☐

= ☐ − ☐

= ☐

2 Use your preferred method to calculate. Show your working.

(a) 788 − 323

(b) 900 − 167

(c) 459 − 366

(d) 558 − 263

3 Use the six digit cards below to form subtraction sentences subtracting a 3-digit number from another 3-digit number. Use your preferred method to work them out.

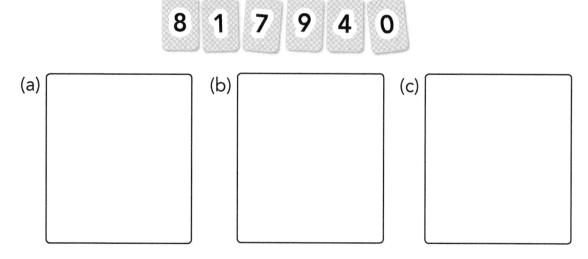

(a)

(b)

(c)

4 Write a number sentence for each question.

(a) The minuend is 429 and the subtrahend is 290. What is the difference?

Answer: _____

(b) What number is 348 less than 695?

Answer: _____

Challenge and extension question

5 The total mass of 1 pack of salt and 1 pack of sugar is 470 grams. The total mass of 2 packs of salt and 1 pack of sugar is 630 grams. What is the mass of 1 pack of salt? What is the mass of 1 pack of sugar?

Answer: _____

7.8 Subtraction with 3-digit numbers (2)

Learning objective Use the column method to subtract 3-digit numbers

Basic questions

1 Use the column method to calculate.

(a) 187 − 31 =

```
  1  8  7
−     3  1
_____

_____
```

(b) 129 − 88 =

```
  1  2  9
−     8  8
_____

_____
```

(c) 151 − 75 =

```
  1  5  1
−     7  5
_____

_____
```

(d) 433 − 265 =

```
  4  3  3
−  2  6  5
_____

_____
```

(e) 800 − 468 =

```
  8  0  0
−  4  6  8
_____

_____
```

(f) 105 − 76 =

```
  1  0  5
−     7  6
_____

_____
```

2 Are these calculations correct? Put a ✓ for yes or a ✗ for no in the box and then make corrections.

(a)
```
     6  4  3
  −  3  0  7
  _____
     3  6  4
```
☐

(b)
```
     4  1  3
  −  3  3  1
  _____
     1  4  2
```
☐

(c)
```
     1  0  5
  −     3  9
  _____
        7  6
```
☐

Corrections:

3 Complete the table.

Minuend	737	562	656	770	312	707
Subtrahend	133	266	475	634	134	668
Difference						

4 Fill in the boxes.

(a)
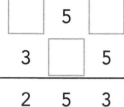

(b)

(c)

5 Application problems.

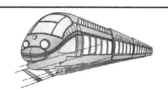

a plane flies at 769 km per hour	a high-speed train travels at 372 km per hour	a car travels at 112 km per hour

(a) How many more kilometres does the plane travel than the high-speed train per hour?

Answer: _____

(b) How many fewer kilometres does the car travel than the high-speed train per hour?

Answer: _____

(c) Using 'how many more' or 'how many fewer', write your own question based on the pictures above, and give the answer.

_____

_____

Challenge and extension question

6 Complete the sentences.

$$
\begin{array}{r}
5\ \bigcirc\ 2 \\
-\quad 4\ \ 3\ \ 9 \\
\hline \\
\hline
\end{array}
$$

(a) If the difference is a 3-digit number, the smallest possible number in the $\bigcirc$ is ☐.

(b) If the difference is a 2-digit number, the greatest possible number in the $\bigcirc$ is ☐.

7.9 Estimating addition and subtraction with 3-digit numbers (1)

 Learning objective Estimate and calculate the answers to addition and subtraction problems

 Basic questions

1 Calculate mentally.

(a) 320 + 80 = ☐   (b) 310 − 90 = ☐

(c) 580 + 100 = ☐   (d) 102 − 29 = ☐

(e) 720 − 60 = ☐   (f) 490 + 60 = ☐

(g) 670 − 600 = ☐   (h) 390 + 200 = ☐

2 Complete the table. One has been done for you.

	232	527	659	707	348	499
The nearest tens number	230					
The nearest hundreds number	200					

3 Estimate to the nearest 10 first and then calculate.

(a) 212 + 168   (b) 438 + 321   (c) 192 + 306

Estimate: ☐   Estimate: ☐   Estimate: ☐

Calculate: Calculate: Calculate:

(d) 626 + 322

Estimate: [ ]

Calculate:

(e) 674 − 418

Estimate: [ ]

Calculate:

(f) 813 − 479

Estimate: [ ]

Calculate:

4 Estimate to the nearest 100 first and then calculate.

(a) 388 + 217

Estimate: [ ]

Calculate:

(b) 657 + 129

Estimate: [ ]

Calculate:

(c) 290 + 426

Estimate: [ ]

Calculate:

(d) 587 + 349

Estimate: [ ]

Calculate:

(e) 477 − 286

Estimate: [ ]

Calculate:

(f) 529 − 380

Estimate: [ ]

Calculate:

5 Application problems.

(a) A school organises 228 Year 3 and 198 Year 4 pupils to go swimming at a local pool. There cannot be more than 450 children in the swimming pool at the same time. Estimate whether all these children can swim at the same time.

Answer: _____

(b) There are 287 apple trees and 343 pear trees in an orchard. Estimate the total number of apple trees and pear trees, and the difference between the two types of tree.

Number of apple and pear trees: _____

Difference: _____

(c) Erin's mum went to a furniture shop with £200. She wanted to buy two rugs for £79 and a set of kitchen chairs for £114.

(i) Estimate whether Erin's mum had enough money for the rugs and chairs she wanted to buy.

Answer: _____

(ii) If she had enough money, how much change should she get? If she did not have enough, how much was she short?

Answer: _____

Challenge and extension question

6 A number consists of three digits: **6**, **4** and **2**.

(a) If you add the number to 350, the result is between 600 and 700. This number is ⬚.

(b) If you subtract 350 from the number, the result is between 100 and 200. This number is ⬚.

7.10 Estimating addition and subtraction with 3-digit numbers (2)

 Learning objective Estimate and calculate the answers to addition and subtraction problems

 Basic questions

1 Calculate mentally.

(a) 450 + 160 = ☐

(b) 660 − 450 = ☐

(c) 370 + 270 = ☐

(d) 610 − 270 = ☐

(e) 480 − 330 = ☐

(f) 530 + 160 = ☐

(g) 360 + 550 = ☐

(h) 1000 − 430 = ☐

2 Estimate first and then calculate.

(a) 431 + 278

To the nearest 10: ☐

To the nearest 100: ☐

Calculate:

(b) 516 + 483

To the nearest 10: ☐

To the nearest 100: ☐

Calculate:

(c) 878 − 356

To the nearest 10: ☐

To the nearest 100: ☐

Calculate:

(d) 623 − 399

To the nearest 10: ☐

To the nearest 100: ☐

Calculate:

3 Calculate with reasoning. Start with the easiest calculation. Think carefully about which one to start with.

(a)   97 + 238 = ☐

(b)   98 + 238 = ☐

(c)   99 + 238 = ☐

(d)   100 + 238 = ☐

(e)   456 + 98 = ☐

(f)   456 + 99 = ☐

(g)   456 + 100 = ☐

(h)   456 + 101 = ☐

(i)   200 − 130 = ☐

(j)   199 − 130 = ☐

(k)   198 − 230 = ☐

(l)   197 − 230 = ☐

(m)   550 − 97 = ☐

(n)   550 − 98 = ☐

(o)   550 − 99 = ☐

(p)   550 − 100 = ☐

4 (a) Estimate and then write the letters of the number sentences.

A. 213 + 176   **B.** 323 + 268   **C.** 334 + 178   **D.** 352 + 278

E. 138 + 363   **F.** 233 + 171   **G.** 268 + 326   **H.** 382 + 156

   (i) Estimating to the nearest 100, the number sentence(s) with a result of:

   400 is/are _____, with result of 500 is/are _____, and with a result of 600 is/are _____.

   (ii) Estimating to the nearest 10, the number sentence(s) with a result of:

   400 is/are _____, with a result of 500 is/are _____, and with a result of 600 is/are _____.

(b) Now calculate.

   (i) 213 + 176 = ☐   (ii) 323 + 268 = ☐

   (iii) 334 + 178 = ☐   (iv) 352 + 278 = ☐

   (v) 138 + 363 = ☐   (vi) 233 + 171 = ☐

   (vii) 268 + 326 = ☐   (viii) 382 + 156 = ☐

5 Use the information in the table to estimate the answers.

Name of building in London	Broadgate Tower	One Churchill Place	Canary Wharf Tower	The Shard
Height (metres)	178	156	244	310

(a) About how many metres higher is The Shard than Broadgate Tower?

Answer: _____

(b) About how many metres shorter is One Churchill Place than Canary Wharf Tower?

Answer: _____

(c) Write two more estimation questions of your own and work out the answers.

_____

_____

 Challenge and extension question

6

123   234   345   789

456   567   678

Choose six of these numbers to write in the boxes. (Think carefully: how many different ways are there of doing this?)

☐ + ☐ = ☐ + ☐ = ☐ + ☐

Chapter 7 test

1 Calculate mentally.

(a) 145 + 70 = ☐

(b) 180 − 18 = ☐

(c) 570 + 300 = ☐

(d) 400 − 180 = ☐

(e) 390 − 66 = ☐

(f) 243 + 52 = ☐

(g) 610 − 280 = ☐

(h) 150 + 68 = ☐

2 Use the column method to calculate.

(a) 217 + 38 = ☐

(b) 384 + 138 = ☐

(c) 157 − 48 = ☐

(d) 500 − 326 = ☐

(e) 108 + 39 = ☐

(f) 101 − 66 = ☐

3 (a) Estimate to the nearest 100 first and then calculate.

(i) 276 + 117

Estimate: [　　]

Calculate:

(ii) 381 + 207

Estimate: [　　]

Calculate:

(iii) 512 − 296

Estimate: [　　]

Calculate:

(iv) 389 − 209

Estimate: [　　]

Calculate:

(b) Estimate to the nearest 10 first and then calculate.

(i) 332 + 176

Estimate: [　　]

Calculate:

(ii) 168 + 223

Estimate: [　　]

Calculate:

(iii) 389 − 277

Estimate: [　　]

Calculate:

(iv) 501 − 137

Estimate: [　　]

Calculate:

4 Fill in the boxes.

(a)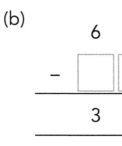

$$\begin{array}{r} \boxed{}\ \boxed{}\ 2 \\ -\ 1\ \ 4\ \boxed{} \\ \hline 5\ \ 5\ \ 0 \end{array}$$

(b)

$$\begin{array}{r} 6\ \ 6\ \boxed{} \\ -\ \boxed{}\ \boxed{}\ 6 \\ \hline 3\ \ 9\ \ 6 \end{array}$$

(c)

$$\begin{array}{r} 8\ \boxed{}\ 0 \\ -\ \boxed{}\ 8\ \boxed{} \\ \hline 3\ \ 1\ \ 8 \end{array}$$

5 Write a number sentence for each question.

(a) The sum of two addends is 560. One addend is 388. What is the other addend?

Number sentence: _____

(b) What number is 39 less than 105?

Number sentence: _____

(c) The subtrahend is 288 and the difference is 109. What is the minuend?

Number sentence: _____

6 A primary school has 337 boys and 368 girls. How many pupils are there altogether?

Answer: _____

7 A large fridge freezer costs £678. Lori's aunt pays the shop assistant £700. How much change should she receive?

Answer: _____

8 In a charity activity, the pupils in Year 2 donated 187 books. The Year 3 pupils donated 226 books. How many more books did Year 3 donate than Year 2?

Answer: _____

9 (a) Samira has collected 295 British postage stamps and 127 other countries' postage stamps. How many stamps has she collected in total?

Answer: _____

(b) Samira collected 250 of these stamps this year and the rest last year. How many stamps did she collect last year?

Answer: _____

10 There are 215 tonnes of sand on a construction site. The amount of cement is 36 tonnes fewer than that of sand and 78 tonnes fewer than that of gravel.

(a) How many tonnes of cement are there on the construction site?

Answer: _____

(b) How much gravel is there?

Answer: _____

Chapter 8 Simple fractions and their addition and subtraction

8.1 Unit fractions and tenths

 Learning objective Use fractions of quantities and count in tenths

Basic questions

1 Write fractions in the boxes to represent the shaded part of each bar.

(a)

(b)

(c)

(d)

(e)

(f)

(g) Which of the fractions above are unit fractions? _____

2 Circle the objects in each diagram to show the fraction given below.

(a)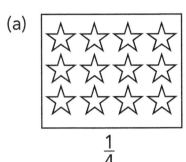

$\frac{1}{4}$

(b)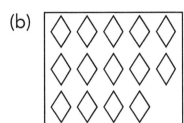

$\frac{1}{7}$

(c)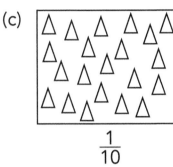

$\frac{1}{10}$

(d)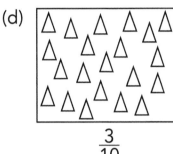

$\frac{3}{10}$

3 Complete the number lines by writing fractions in the boxes.

(a)
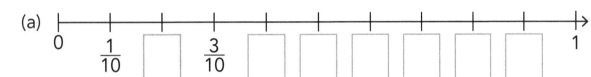

0    $\frac{1}{10}$ ☐ $\frac{3}{10}$ ☐ ☐ ☐ ☐ ☐   1

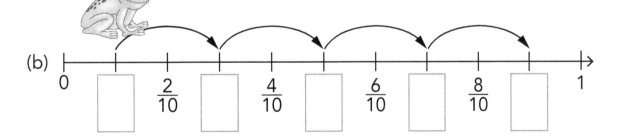

(b)

0   ☐ $\frac{2}{10}$ ☐ $\frac{4}{10}$ ☐ $\frac{6}{10}$ ☐ $\frac{8}{10}$ ☐   1

Each unit on the number line is ☐.

 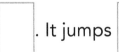 starts to jump from ☐. It jumps ☐ units each time and finally lands on ☐.

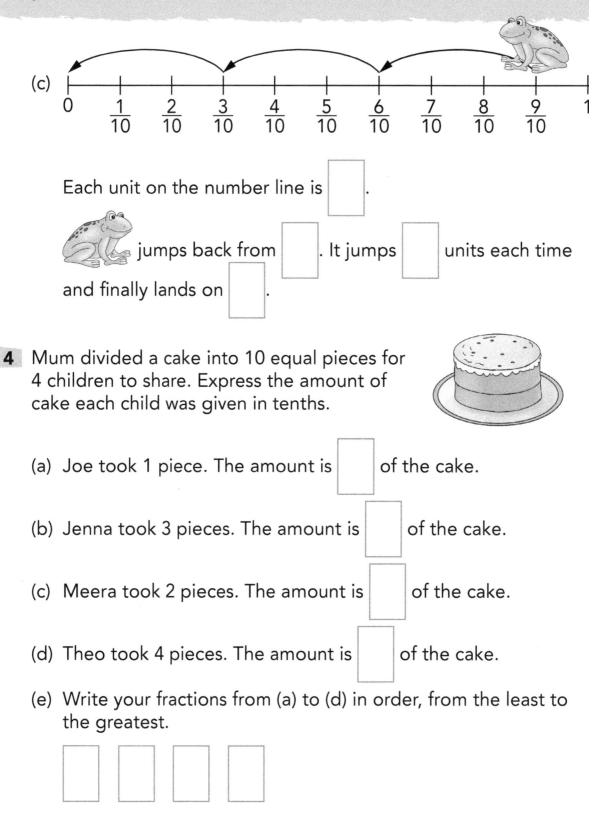

(c)

Each unit on the number line is ☐.

jumps back from ☐. It jumps ☐ units each time

and finally lands on ☐.

4 Mum divided a cake into 10 equal pieces for 4 children to share. Express the amount of cake each child was given in tenths.

(a) Joe took 1 piece. The amount is ☐ of the cake.

(b) Jenna took 3 pieces. The amount is ☐ of the cake.

(c) Meera took 2 pieces. The amount is ☐ of the cake.

(d) Theo took 4 pieces. The amount is ☐ of the cake.

(e) Write your fractions from (a) to (d) in order, from the least to the greatest.

☐ ☐ ☐ ☐

5 Write the unit fractions and tenths in order, from the least to the greatest. (Hint: you may use a number line to help you.)

$$\frac{1}{10} \qquad \frac{9}{10} \qquad \frac{1}{2} \qquad \frac{7}{10} \qquad \frac{1}{5} \qquad \frac{3}{10} \qquad \frac{1}{4}$$

☐ ☐ ☐ ☐ ☐ ☐ ☐

8.2 Non-unit fractions

Learning objective Use non-unit fractions of quantities

Basic questions

1 Count the items in the pictures and write fractions in the boxes.

(a)

The number of apples is ☐ of the total.

The number of strawberries is ☐ of the total.

The number of apples and bananas is ☐ of the total.

(b)
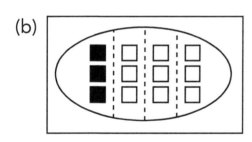

The number of black squares is ☐ of the total.

The number of white squares is ☐ of the total.

(c) Which of the fractions in (a) and (b) are unit fractions?

☐

Which of the fractions in (a) and (b) are non-unit fractions?

☐

2 Write fractions in the boxes to represent the shaded part of each diagram.

(a)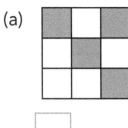

(b)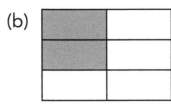

(c)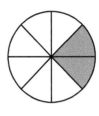

☐   ☐   ☐ or ☐

3 Does each fraction below represent the shaded part of the whole correctly? Put a ✓ for yes or a ✗ for no in the box.

(a)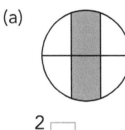

$\frac{2}{6}$ ☐

(b)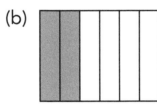

$\frac{1}{3}$ ☐

(c)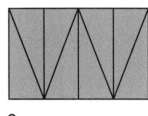

$\frac{8}{8}$ ☐

4 Fill in the boxes.

(a) A strip of ribbon, with a length of 1 metre, is cut into 7 equal pieces. Each strip is ☐ metres long. 4 pieces are ☐ metres long.

(b) (i) $\frac{2}{15}$ of 15 ▲ is ☐ ▲.   (ii) $\frac{1}{5}$ of 15 ▲ is ☐ ▲.

(iii) $\frac{2}{5}$ of 15 ▲ is ☐ ▲.   (iv) $\frac{3}{5}$ of 15 ▲ is ☐ ▲.

5 Write the numbers in order, from the least to the greatest.

(a) $\frac{1}{7}$ $\frac{5}{7}$ $\frac{6}{7}$ $\frac{4}{7}$ $\frac{2}{7}$ ☐ ☐ ☐ ☐ ☐

(b) $\frac{4}{9}$ $\frac{2}{9}$ $\frac{8}{9}$ $\frac{7}{9}$ 1   ☐ ☐ ☐ ☐ ☐

Challenge and extension question

6 Calculate and fill in the boxes. (Hint: you may draw diagrams to help you find the answers.)

(a) $\frac{2}{18}$ of 36 ▲ is ▢ ▲.

(b) $\frac{2}{12}$ of 36 ▲ is ▢ ▲.

(c) $\frac{2}{9}$ of 36 ▲ is ▢ ▲.

(d) $\frac{2}{6}$ of 36 ▲ is ▢ ▲.

(e) $\frac{2}{4}$ of 36 ▲ is ▢ ▲.

(f) $\frac{2}{3}$ of 36 ▲ is ▢ ▲.

8.3 Equivalent fractions

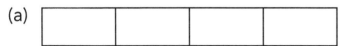 **Learning objective** Recognise and show equivalent
fractions

 Basic questions

1 Look at the diagram and then fill in the boxes.

(a)

(i) The rectangle is divided into 4 equal parts. What fraction
of the rectangle does each part represent?

(ii) What fraction of the rectangle does 4 parts represent?

(iii) What whole number is this equal to?

(b) What is 6 one-sixths as a fraction?

This is equal to 1.

2 Look at each picture and write two different fractions in the boxes.
You may discuss the answers with your friends.

(a)

☐ or ☐

(b)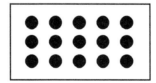

☐ or ☐

(c)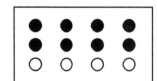

☐ or ☐

3 Fill in the boxes.

(a) 10 one-tenths of a pound are $\dfrac{\boxed{}}{\boxed{}}$.

This is equivalent to $\boxed{}$ pounds.

(b) Write >, < or = in each $\bigcirc$.

(i) $\dfrac{1}{12} \bigcirc \dfrac{1}{9}$　　　　(ii) $\dfrac{4}{10} \bigcirc \dfrac{2}{10}$　　　　(iii) $\dfrac{3}{3} \bigcirc \dfrac{9}{9}$

(c) 4 one-fourths are $\dfrac{\boxed{}}{\boxed{}}$. This is equivalent to $\boxed{}$.

(d) $\dfrac{1}{4}$ of 8 chocolates is $\boxed{}$ chocolates.

(e) 4 one-sixths are $\dfrac{\boxed{}}{\boxed{}}$. 5 $\dfrac{\boxed{}}{\boxed{}}$ are $\dfrac{5}{5}$ and this is equivalent to $\boxed{}$.

(f) Mr Lee has 3 cockerels and 5 hens. The number of hens is $\dfrac{\boxed{}}{\boxed{}}$ the total number of cockerels and hens.

4 Circle equivalent fractions and then draw lines to match them.

$\dfrac{5}{6}$　　$\dfrac{2}{7}$　　$\dfrac{3}{5}$　　$\dfrac{4}{6}$

$\dfrac{3}{4}$

$\dfrac{1}{2}$

$\dfrac{4}{14}$

$\dfrac{9}{12}$

$\dfrac{2}{3}$

$\dfrac{6}{10}$　　$\dfrac{3}{6}$　　$\dfrac{10}{12}$

5 Write a fraction to represent the shaded part in each diagram.

(a)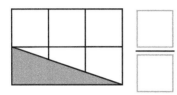

(b)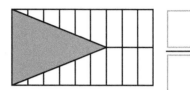

6 Think carefully about how to answer these.

(a) If 48 bottles of water are divided into 8 equal parts, each part has ☐ bottles of water. If they are divided into 6 equal parts, each part has ☐ bottles of water.

(b) If Lily takes away $\frac{3}{8}$ of the water and Ellis takes away $\frac{2}{6}$, then _____ takes more.

(c) If Holly wants to take $\frac{1}{5}$, is it possible? Explain why/why not.

_____

(d) What fraction of water can Holly take? Explain why.

_____

8.4 Addition and subtraction of simple fractions

 Learning objective Add and subtract fractions with the same denominator

 Basic questions

1 Look at the diagrams and then add the fractions.

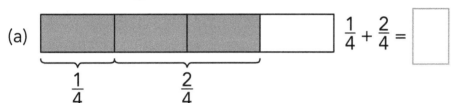

(a) $\dfrac{1}{4} + \dfrac{2}{4} =$ ☐

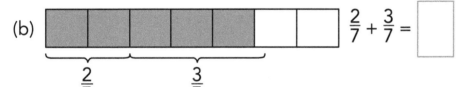

(b) $\dfrac{2}{7} + \dfrac{3}{7} =$ ☐

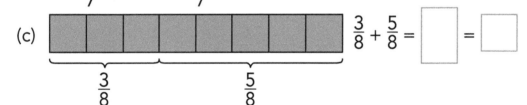

(c) $\dfrac{3}{8} + \dfrac{5}{8} =$ ☐ $=$ ☐

2 Look at the diagrams and complete the subtraction calculations.

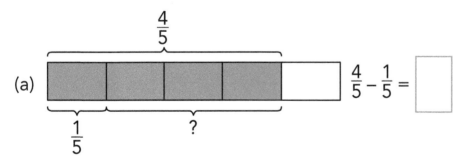

(a) $\dfrac{4}{5} - \dfrac{1}{5} =$ ☐

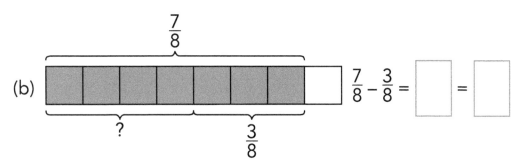

(b) $\dfrac{7}{8} - \dfrac{3}{8} = \boxed{} = \boxed{}$

3 Add the fractions.

(a) $\dfrac{1}{2} + \dfrac{1}{2} = \boxed{}$

(b) $\dfrac{2}{7} + \dfrac{4}{7} = \boxed{}$

(c) $\dfrac{1}{4} + \dfrac{3}{4} = \boxed{}$

(d) $\dfrac{1}{3} + \dfrac{1}{3} = \boxed{}$

(e) $\dfrac{5}{9} + \dfrac{2}{9} = \boxed{}$

(f) $\dfrac{3}{5} + \dfrac{2}{5} = \boxed{}$

4 Subtract the fractions.

(a) $\dfrac{5}{8} - \dfrac{3}{8} = \boxed{}$

(b) $1 - \dfrac{1}{2} = \boxed{}$

(c) $\dfrac{4}{9} - \dfrac{4}{9} = \boxed{}$

(d) $\dfrac{6}{7} - \dfrac{1}{7} = \boxed{}$

(e) $\dfrac{3}{4} - \dfrac{1}{4} = \boxed{}$

(f) $\dfrac{4}{5} - 0 = \boxed{}$

5 Anna was reading a book. The book had 24 pages in total. She read 3 pages on the first day, 4 pages on the second day and 5 pages on the third day.

(a) Write the fraction of the book Anna read on each day.

First day: $\boxed{}$

Second day: $\boxed{}$

Third day: $\boxed{}$

(b) What fraction of the book did Anna read on the

   first two days? ☐

(c) What fraction of the book had Anna read after three days? ☐

(d) What fraction of the book had Anna not read? ☐

   How many pages had she not read? ☐

Challenge and extension question

6 Maya's grandma bought 12 pears to share. She gave Maya and her brother Theo 3 pears each. She then kept 1 pear to herself and gave the rest to the children's mum.

(a) Complete the sentence by writing the correct fractions in the boxes.

   Grandma got ☐ of all the pears; Maya got ☐ of all the pears;

   Theo got ☐ of all the pears; and Mum got ☐ of all the pears.

(b) What fraction of all the pears did Maya and Theo get in total? Write the number statement and calculate the answer.

Number statement: _____

(c) What fraction of all the pears did Maya, Theo and Mum get altogether?

Number statement: _____

(d) Who got the most pears? Who got the least? What is the difference? Write the number statement and find the difference.

Answer: _____ got the most; _____ got the least.

The difference is: _____ = _____ of the total number of pears.

Express the difference as a whole number. It is ☐ pears.

1 Fill in the boxes.

(a) If 10 eggs are shared by 10 children equally, one child gets $\boxed{}$ of all the eggs. Five children get $\boxed{}$ or $\boxed{}$ of all the eggs.

(b) $1 = \dfrac{\boxed{}}{2} = \dfrac{10}{\boxed{}} = \dfrac{\boxed{}}{15}$.

(c) 4 _____ is $\dfrac{4}{6}$. It is read as _____.

Seven-eighths is written as $\boxed{}$.

(d) $\dfrac{2}{5}$ of the 10 ▲ are $\boxed{}$ ▲. $\dfrac{1}{3}$ of $\boxed{}$ ★ is 4 ★.

(e) The fewer the number of parts the same whole is equally divided into, the _____ each part becomes.

The _____ the number of parts the same whole is equally divided into, the smaller each part becomes.

2 Compare the fractions using the diagrams to help you. Write >, < or = in each $\bigcirc$.

(a)

$\dfrac{2}{2} \bigcirc \dfrac{4}{4}$

(b)

$\dfrac{1}{12} \bigcirc \dfrac{1}{8}$

(c)

$\dfrac{4}{5} \bigcirc \dfrac{2}{5}$

3 Draw lines to match the equivalent fractions.

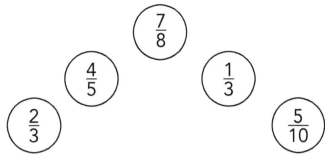

4 Write the fractions in order, from the least to the greatest.

(a) $\frac{1}{5}$ $\frac{1}{12}$ $\frac{1}{9}$ $\frac{1}{2}$ 1

(b) $\frac{5}{7}$ $\frac{4}{7}$ $\frac{2}{7}$ $\frac{6}{7}$ 1

5 Work out these calculations.

(a) $\frac{1}{4} + \frac{1}{4} =$ ☐

(b) $\frac{7}{9} - \frac{2}{9} =$ ☐

(c) $\frac{3}{10} + \frac{7}{10} =$ ☐

(d) $\frac{2}{3} - \frac{1}{3} =$ ☐

(e) $\frac{1}{11} + \frac{8}{11} =$ ☐

(f) $\frac{1}{5} + \frac{4}{5} =$ ☐

(g) $\frac{7}{10} - \frac{3}{10} =$ ☐

(h) $\frac{1}{5} - \frac{1}{5} =$ ☐

(i) $1 - \frac{7}{8} =$ ☐

6 Look at the diagram and answer the questions.

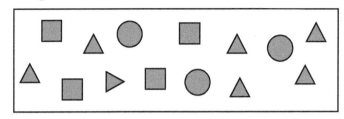

(a) There are ☐ shapes altogether.

(b) Out of the total number of shapes, the number

of ⬤ is $\dfrac{}{}$, the number of ◼ is $\dfrac{}{}$,

and the number of ▲ is $\dfrac{}{}$.

(c) What fraction of all the shapes is the number of ◼ and ▲ altogether? Write the number sentence and calculate.

Number sentence: _____

(d) What shape makes up the greatest fraction of the total number of shapes?

What shape makes up the least? What is the difference?

Answer: _____ makes up the greatest fraction of all the

shapes; _____ makes up the least.

The difference is: _____

7 12 toys were given to two boys and two girls. Amy got 3 toys, Joe got 4 toys, Ravi got 2 toys and Min got the rest. Use your knowledge of fractions to answer the questions.

(a) What fraction of the toys did Amy get?

_____

(b) What fraction of the toys did Joe and Ravi get altogether?

_____

(c) Did the boys and girls get the same quantity of toys?

_____

8 Have you played Tangram? It is a puzzle consisting of 7 flat shapes, called tans. The 7 shapes can be put together in different ways. You can make thousands of beautiful designs with this simple 7-piece magic puzzle! Let's do some fraction problems using the Tangram.

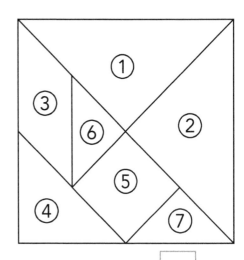

(a) What fraction of the square does Shape ① take up? ☐.

(b) What fraction of the square does Shape ④ take up? ☐.

(c) What fraction of the square does Shape ⑦ take up? ☐.

(d) What fraction of the square does Shape ⑤ take up? ☐.

(e) What fraction of the square does Shape ③ take up? ☐.

(f) Write a question of your own and give the answer.

_____

_____

_____

Chapter 9 Multiplying and dividing by a 1-digit number

9.1 Multiplying by whole tens and hundreds (1)

 Learning objective Multiply tens and hundreds by 1-digit numbers

 Basic questions

1 Calculate with reasoning.

(a) $4 \times 7 =$ [ ] (b) $5 \times 3 =$ [ ] (c) $6 \times 9 =$ [ ]

(d) $4 \times 70 =$ [ ] (e) $50 \times 3 =$ [ ] (f) $60 \times 9 =$ [ ]

(g) $4 \times 700 =$ [ ] (h) $500 \times 3 =$ [ ] (i) $6 \times 90 =$ [ ]

2 Fill in the boxes.

(a) (i) $60 + 60 + 60 + 60 =$ [ ]

[ ] $\times$ [ ] $=$ [ ]

(ii) $200 + 200 + 200 + 200 + 200 =$ [ ]

[ ] $\times$ [ ] $=$ [ ]

(b) When calculating 6×30 mentally, you can think of it as

6 multiplied by [ ] tens, which is [ ] tens.

So the answer to 6×30 is [ ].

(c) There are [ ] zeros at the end of the product of 700×9.

3 Write >, < or = in each ◯.

(a) 9×50 ◯ 90×5

(b) 40×2 ◯ 3×20

(c) 8×200 ◯ 500×6

(d) 300×3 ◯ 5×200

(e) 7×90 ◯ 6×600

(f) 600×4 ◯ 4×600

4 Draw lines to match each calculation to the correct answer.

(a) 5×70 3500

(b) 500×7 35

(c) 5×7000 350

(d) 5×7 35 000

5 Fill in the boxes.

(a) [ ] $\times 3 = 1200$

(b) $700 \times$ [ ] $= 1400$

(c) $2000 =$ [ ] $\times 400$

(d) $3200 = 800 \times$ [ ]

6 Write a number sentence for each question.

(a) What is the sum of 40 nines?

Number sentence: _____

(b) What is 3 times 600?

Number sentence: _____

(c) A number is divided by 400 and both its quotient and the remainder are 8. What is the number?

Number sentence: _____

7 A pack of salt weighs 500 grams.
How many kilograms do 8 packs of salt
weigh? (Note: 1 kilogram = 1000 grams)

Answer: _____

 Challenge and extension question

8 One plastic box contains 50 pens. 8 plastic boxes can fit into one
large case. The price of each pen is £3. How much will two large
cases of pens cost?

Answer: _____

9.2 Multiplying by whole tens and hundreds (2)

 Learning objective Multiply by multiples of ten and hundred

 Basic questions

1 Calculate with reasoning.

(a) $3 \times 9 = $ ☐

(b) $4 \times 8 = $ ☐

(c) $7 \times 6 = $ ☐

(d) $3 \times 90 = $ ☐

(e) $40 \times 8 = $ ☐

(f) $700 \times 6 = $ ☐

(g) $3 \times 900 = $ ☐

(h) $4 \times 80 = $ ☐

(i) $7 \times 600 = $ ☐

(j) $30 \times 90 = $ ☐

(k) $40 \times 80 = $ ☐

(l) $70 \times 60 = $ ☐

2 Fill in the boxes.

(a) When calculating 7×500 mentally, you can think of it as 7 multiplied by ☐ hundreds, which is ☐ hundreds. So the answer to 7×500 is ☐.

(b) There are ☐ zeros at the end of the product of 500×6.

3 Write >, < or = in each ◯.

(a) 6 × 50 ◯ 60 × 5

(b) 400 × 5 ◯ 40 × 50

(c) 7 × 700 ◯ 900 × 5

(d) 60 × 30 ◯ 300 × 6

(e) 7 × 800 ◯ 50 × 10

(f) 900 × 2 ◯ 3 × 600

4 Draw lines to match each calculation with the correct answer.

(a) 5 × 80　　　　　　　　　　　4000

(b) 500 × 8　　　　　　　　　　40

(c) 50 × 800　　　　　　　　　400

(d) 5 × 8　　　　　　　　　　40 000

5 Multiple choice questions.

(a) If ▲ = 3, ■ = 600, ● = 80, then ■ − ▲ × ● = ☐.

　A. 1800　　B. 240　　C. 360　　D. 300

(b) If 36 = ★ + ★ + ★ + ★, then 70 × ★ = ☐.

　A. 280　　B. 630　　C. 560　　D. 490

6 Application problems.

(a) A farm has 400 hens. The number of chicks is 4 times the number of hens. How many chicks are there? How many chicks and hens are there in total?

Number of chicks: _____

Total number of chicks and hens: _____

(b) (i) A washing machine costs £300. A laptop computer is 5 times as expensive as the washing machine. How much is a laptop computer?

Answer: _____

(ii) How much more expensive is a laptop than a washing machine?

Answer: _____

Challenge and extension question

7 Fill in the boxes.

$2000 \times 8 = 200 \times$ ⬜ $=$ ⬜ $\times 800 =$ ⬜ $\times 4 = 1000 \times$

⬜ $= 100 \times$ ⬜ $=$ ⬜ $\times 40 = 500 \times$ ⬜

9.3 Writing number sentences

Learning objective Write number sentences for multiplication problems

Basic questions

1 Write a number sentence for each calculation.

(a) There are 6 cupcakes in one box.
How many cupcakes are there in 11 boxes?

Number sentence: _____

(b) Evie and 3 friends are going on a trip. Everyone needs to pay £30 for the trip. How much do they pay in total?

Number sentence: _____

(c) There are 4 packets of nuts in one box. Each pack costs £12. How much will 3 packs of nuts cost?

Number sentence: _____

2 Look at the pictures. Draw a line to match each question to the correct number sentence.

	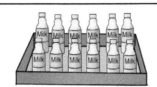	
BAKED BEANS		
12 cans per box	12 bottles per box	10 kilograms per sack
£70 per box	£45 per box	£50 per sack

(a) How much do 2 boxes of baked beans cost?   4 × 10

(b) How much do 2 boxes of milk cost?   2 × 12

(c) How many bottles of milk are there in 2 boxes?   2 × 70

(d) What is the weight of 4 sacks of rice?   2 × 45

 Challenge and extension questions

3 (a) Draw a line to match each calculation to the correct answer.

        39 × 8   26 × 3   92 × 4

        368   312   276   78

(b) The number that does not match a calculation is [   ].

    Can you write a calculation to go with it? _____

4 Write a suitable condition.
Then write a number sentence
and calculate the answer.

(a) A box can be filled up with 8 pieces of chocolate. _____

_____.

(b) How many boxes can be filled up with _____?

5 Some of the two whole numbers with a sum of 16 are: 0 and 16, 1
and 15, 2 and 14, 3 and 13, 4 and 12. List the others.

_____

_____

_____

_____

The two numbers whose product is the greatest are ☐ and ☐.

9.4 Multiplying a 2-digit number by a 1-digit number (1)

 Learning objective Multiply 2-digit numbers by 1-digit numbers

 Basic questions

1 Look at the array of stars, then complete the multiplication and addition calculations.

> How many ☆ are there altogether?

> Split 13 into 10 and 3 first and multiply each by 6. Then add the two products.

> I use multiplication to find out.
>
> 13 × 6 = ?

First multiply:

☐ × ☐ = ☐   ☐ × ☐ = ☐

then add:

☐ + ☐ = ☐

therefore:

13 × 6 = ☐

2 Work out these calculations.

(a) 7 × 74 = ☐

(b) 26 × 8 = ☐

(c) 64 × 9 = ☐

(d) 2 × 93 = ☐

(e) 89 × 6 = ☐

(f) 5 × 54 = ☐

3 Application problems.

(a) A group of Year 3 pupils are gathered in a school sports hall. Each row has 15 pupils and there are 8 rows. How many pupils are there altogether?

Answer: _____

(b) In a school, 24 pupils joined the football team. The number of pupils that joined the choir is twice the number that joined the football team. How many pupils joined the choir?

Answer: _____

Challenge and extension question

4 Write < or > in each ◯.

(a) 14 × 7 ◯ 17 × 4   (b) 27 × 3 ◯ 23 × 7

(c) 19 × 5 ◯ 15 × 9   (d) 45 × 2 ◯ 42 × 5

What did you notice?

_____

_____

9.5 Multiplying a 2-digit number by a 1-digit number (2)

Learning objective Multiply 2-digit numbers by 1-digit numbers

Basic questions

1 What is the total cost of 3 model sailing boats?

£81   £81   £81

$3 \times 81 = ?$

Method 1:

(a)

```
      8   1
  ×       3
  ┌───┬───┬───┐
  │   │   │   │
  └───┴───┴───┘
```

Usually, the number with more digits is placed on the top row.

Method 2:

(b)

```
      8   1
  ×       3
  ─────────────
          3    ...  □ × □
  2   4   0    ...  □ × □
  ─────────────
  2   4   3
  ─────────────
```

The total cost of 3 model sailing boats is £ ☐ .

2 One model car costs £49.
How much will two model cars cost?

49 × 2 = ?

(a)

```
      4   9
  ×       2
  _____
      |   |   |
  _____
```

(b) Check your answer: 2 × 49 = ?

2 × 40 = ☐

2 × 9 = ☐

☐ + ☐ = ☐

Think: Is there another method to check your answer?

3 Use the column method to calculate. Don't forget to check your work with your preferred method.

(a) 4 × 62 =

(b) 32 × 2 =

(c) 17 × 5 =

(d) 7 × 51 =

4 Are these calculations correct? Put a ✓ for yes and a ✗ for no in the box and then make corrections.

(a)
```
      1   4
  ×       4
  ─────────
      4   6
  ─────────
```
□

Corrections:

(b)
```
          5   3
  ×           3
  ─────────────
      1   5   9
  ─────────────
```
□

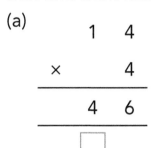

Challenge and extension question

5 Write suitable numbers in the boxes to make the calculation correct.

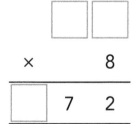

```
    □   □
  ×       8
  ─────────
  □   7   2
```

9.6 Multiplying a 2-digit number by a 1-digit number (3)

Learning objective Multiply 2-digit numbers by 1-digit numbers

 Basic questions

1 Calculate mentally.

(a) $2 \times 8 + 4 =$ ☐

(b) $9 \times 3 + 3 =$ ☐

(c) $6 \times 8 + 5 =$ ☐

(d) $7 \times 7 + 4 =$ ☐

(e) $8 \times 7 + 5 =$ ☐

(f) $6 \times 6 + 5 =$ ☐

(g) $9 \times 5 + 7 =$ ☐

(h) $0 \times 8 + 3 =$ ☐

2 Use the column method to calculate.

(a) $77 \times 4 =$ ☐

(b) $9 \times 36 =$ ☐

(c) $25 \times 8 =$ ☐

(d) $4 \times 75 =$ ☐

3 First estimate, and then calculate using the column method.

(a) 49 × 7 = ▢

    I estimate the product

    is between ▢ and ▢ .

Calculate:

(b) 75 × 6 = ▢

    I estimate the product

    is between ▢ and ▢ .

Calculate:

(c) 88 × 8 = ▢

    I estimate the product

    is between ▢ and ▢ .

Calculate:

4 Are these calculations correct? Put a ✓ for yes and a ✗ for no in the box and then make corrections.

(a)

```
      9  9
×        9
─────────
   8  8  1
```
▢

(b)

```
      2  6
×        8
─────────
   1  9  6
```
▢

(c)

```
      6  8
×        5
─────────
   3  0  0
```
▢

Corrections:

5 Draw a line to match each calculation with the correct answer.

$$25 \times 7 = \qquad 56 \times 4 = \qquad 9 \times 42 = \qquad 5 \times 63 =$$

315   378   224   175

6 Application problems.

(a) One music player costs £98. How much do 4 music players cost?

(b) One model sailing boat costs £57. How much do 6 boats cost?

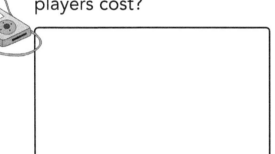

Challenge and extension question

7 Write suitable numbers in the boxes to make the calculation correct.

$$
\begin{array}{r}
1\ \square\ \square \\
\times \qquad\ \ 8 \\
\hline
\square\ 5\ \square\ 0
\end{array}
$$

9.7 Multiplying a 3-digit number by a 1-digit number (1)

 Learning objective Multiply 3-digit numbers by 1-digit numbers

 Basic questions

1 Calculate mentally.

(a) $3 \times 800 = \boxed{}$

(b) $3 \times 80 = \boxed{}$

(c) $3 \times 8 = \boxed{}$

(d) $20 \times 8 = \boxed{}$

(e) $200 \times 8 = \boxed{}$

(f) $2 \times 8 = \boxed{}$

(g) $900 \times 7 = \boxed{}$

(h) $90 \times 7 = \boxed{}$

(i) $9 \times 7 = \boxed{}$

(j) $60 \times 6 = \boxed{}$

(k) $700 \times 4 = \boxed{}$

(l) $3 \times 9 = \boxed{}$

2 Split the numbers into hundreds, tens and ones. One has been done for you.

(a) $316 = 300 + 10 + 6$

(b) $427 = \boxed{} + \boxed{} + \boxed{}$

(c) $987 = \boxed{} + \boxed{} + \boxed{}$

(d) $634 = \boxed{} + \boxed{} + \boxed{}$

3 Work out these calculations. Show your working.

(a) 3 × 316 = [ ]

3 × [ ] = [ ]

3 × [ ] = [ ]

[ ] + [ ] + [ ] = [ ]

(b) 3 × 316 = [ ]

$$\begin{array}{r} 3 \quad 1 \quad 6 \\ \times \qquad\qquad 3 \\ \hline \\ \hline \end{array}$$

4 Work out these calculations. Show your working.

(a) 427 × 4 = [ ]

(b) 634 × 6 = [ ]

(c) 370 × 5 = [ ]

5 Use the column method to calculate.

(a) 171 × 7 = [ ]

(b) 3 × 279 = [ ]

(c) 6 × 309 = [ ]

6 Are these calculations correct? Put ✓ for yes and ✗ for no in the box and then make corrections.

(a)
```
      2   3   5
  ×           4
  —————————————
      9   4   0
```
☐

(b)
```
      1   0   7
  ×           5
  —————————————
      8   3   5
```
☐

(c)
```
      6   7   2
  ×           8
  —————————————
  5   3   7   6
```
☐

Corrections:

Challenge and extension question

7 Think carefully to complete these calculations.

(a) Fill in the boxes with suitable numbers.

(i)
```
      4   1   6
  ×           7
  —————————————
  ☐   ☐   ☐   2
```

(ii)
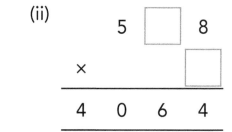
```
      5   ☐   8
  ×           ☐
  —————————————
  4   0   6   4
```

(b) Work out the number that each shape stands for.

```
      ■   ▲   ●   4
  ×               3
  —————————————————
      5   ■   ▲   ●
```

■ = ☐

▲ = ☐

● = ☐

9.8 Multiplying a 3-digit number by a 1-digit number (2)

Learning objective Multiply 3-digit numbers by 1-digit numbers

Basic questions

1 Calculate mentally.

(a) $5 \times 5 =$ ☐

(b) $50 \times 5 =$ ☐

(c) $500 \times 5 =$ ☐

(d) $2 \times 5 =$ ☐

(e) $20 \times 5 =$ ☐

(f) $200 \times 5 =$ ☐

(g) $3 \times 7 =$ ☐

(h) $30 \times 7 =$ ☐

(i) $300 \times 7 =$ ☐

(j) $9 \times 4 =$ ☐

(k) $90 \times 4 =$ ☐

(l) $900 \times 4 =$ ☐

2 Work out the calculations and complete the sentences.

(a)
```
    3  6  0
×         4
─────────────
☐  ☐  ☐  ☐
```
Multiplying 360 ones by 4 is ☐ ones.

(b)
```
    3  6  0
×         4
─────────────
☐  ☐  ☐  ☐
```
Multiplying 36 tens by 4 is ☐ tens.

3 Use the column method to calculate. (First estimate how many digits the product will have.)

(a) 130 × 6 = ☐

(b) 450 × 9 = ☐

(c) 8 × 250 = ☐

Did you estimate correctly?

If you multiply a 3-digit number by a 1-digit number, the product

will be a ☐ -digit number or a ☐ -digit number.

Don't forget the zero(s) at the end of the number!

4 (a) There are ☐ zeros at the end of the product 4 × 5 × 5.

(b) There are ☐ zeros at the end of the product 125 × 8 × 10.

5 Application problems.

(a) Zainab finished reading a book in 9 days. She read 120 pages a day. How many pages are there in the book?

Answer: _____

(b) There are 205 cows on a farm. There are 4 times as many sheep as there are cows. How many sheep are there?

Answer: _____

Challenge and extension questions

6 Multiple choice question.

$$24 \times \blacktriangle = 2400 \times \bullet.$$

Comparing the relationship between $\blacktriangle$ and $\bullet$, the conclusion is $\boxed{}$.

    A. $10 \times \blacktriangle = \bullet$ **B.** $100 \times \bullet = \blacktriangle$ **C.** $100 \times \blacktriangle = \bullet$

    D. $\blacktriangle = \bullet$ **E.** $10 \times \bullet = \blacktriangle$

7 Look at the table. First, work out the product of the numbers in rows A and B. Then compare how the product has changed in relation to one of the other numbers multiplied. What did you find?

A	400	400	800	800
B	3	6	3	6
A × B				

My findings:  _____

_____

9.9 Practice and exercise

Learning objective Solve multiplication problems

Basic questions

1 Calculate mentally.

(a) $9 \times 30 =$ ☐

(b) $13 \times 7 =$ ☐

(c) $2 \times 30 + 3 =$ ☐

(d) $800 \times 5 =$ ☐

(e) $16 \times 5 =$ ☐

(f) $3 \times 30 - 27 =$ ☐

2 Use the column method to calculate.

(a) $7 \times 42 =$ ☐

(b) $6 \times 134 =$ ☐

(c) $480 \times 7 =$ ☐

3 Write >, < or = in each ◯.

(a) 604×0 ◯ 236

(b) 36×6 ◯ 214

(c) $37 + 3$ ◯ 37×3

(d) 28×8 ◯ 88×2

(e) 123×4 ◯ 124×3

(f) 240×5 ◯ 1200

4 Complete the sentences.

(a) There are ☐ zeros at the end of the product 2500 × 8.

(b) The product of a 1-digit number and a 3-digit number is
     a ☐ -digit number or a ☐ -digit number.

(c) The product of 49 × 5 is a ☐ -digit number.

     The product is between ☐ and ☐ .

5 Write a number sentence for each question.

(a) What is the product of 650 multiplied by 7?

Number sentence: _____

(b) What is the sum of 4 lots of seven hundred and twenty-threes?

Number sentence: _____

(c) What is 5 times 106?

Number sentence: _____

(d) Find the product of 38 and 8.

Number sentence: _____

6 Finn's weight is 39 kilograms. His father's weight is twice the weight
of Finn's. What is his father's weight?

Answer: _____

7 Ella was reading a book. She read 18 pages per day for 9 days and still had 42 pages left. How many pages were there in the book?

Answer: _____

 Challenge and extension questions

8 Some bookshelves were delivered to a school. Each bookshelf has 4 shelves. Mrs Lee put 26 books on each shelf. How many books did she put on 3 of the bookshelves?

Answer: _____

9 Fill in the boxes.

(a)
```
    7 □ 8
  ×     □
  ───────
  □ □ 4 0
```

(b)
```
    1 9 □
  ×     □
  ───────
    □ 6 0
```

9.10 Dividing whole tens and whole hundreds

 Learning objective Use division facts to divide multiples of 10 and 100

 Basic questions

1 Look at these division facts.

$18 \div 6 = 3$	$24 \div 3 = 8$	$28 \div 4 = 7$	$6 \div 2 = 3$
$180 \div 6 = 30$	$240 \div 3 = 80$	$280 \div 4 = 70$	$600 \div 2 = 300$

(a) Now work out these division facts.

(i) $8 \div 4 = $ ☐   (ii) $16 \div 4 = $ ☐

(iii) $32 \div 4 = $ ☐   (iv) $20 \div 4 = $ ☐

(v) $80 \div 4 = $ ☐   (vi) $160 \div 4 = $ ☐

(vii) $320 \div 4 = $ ☐   (viii) $200 \div 4 = $ ☐

(b) Try these.

(i) $9 \div 3 = $ ☐   (ii) $8 \div 2 = $ ☐

(iii) $5 \div 5 = $ ☐   (iv) $10 \div 5 = $ ☐

(v) $90 \div 3 = $ ☐   (vi) $80 \div 2 = $ ☐

(vii) $50 \div 5 = $ ☐   (viii) $100 \div 5 = $ ☐

(ix) $900 \div 3 = $ ☐   (x) $800 \div 2 = $ ☐

(xi) $500 \div 5 = $ ☐   (xii) $1000 \div 5 = $ ☐

Multiplying and dividing by a 1-digit number

2 Division calculations can be worked out using multiplication facts. Complete these calculations.

(a) ☐ × 60 = 480

(b) 480 ÷ 60 = ☐

(c) 480 ÷ ☐ = 60

(d) ☐ × 70 = 560

(e) 560 ÷ 70 = ☐

(f) 560 ÷ ☐ = 70

(g) 80 × ☐ = 720

(h) 720 ÷ 80 = ☐

(i) 720 ÷ ☐ = 80

3 Work out these calculations using your preferred method.

(a) 400 ÷ 50 = ☐

(b) 210 ÷ 7 = ☐

(c) 540 ÷ 90 = ☐

(d) 40 ÷ 2 = ☐

(e) 400 ÷ 8 = ☐

(f) 210 ÷ 30 = ☐

(g) 540 ÷ 6 = ☐

(h) 40 ÷ 20 = ☐

(i) 120 ÷ 30 = ☐

(j) 250 ÷ 50 = ☐

(k) 150 ÷ 30 = ☐

(l) 810 ÷ 90 = ☐

(m) 360 ÷ 60 = ☐

(n) 630 ÷ 90 = ☐

(o) 490 ÷ 70 = ☐

(p) 450 ÷ 90 = ☐

4 Write a number sentence for each question.

(a) The dividend is 270 and the divisor is 3. What is the quotient?

Number sentence: _____

(b) How many times 50 is 350?

Number sentence: _____

(c) Divide 640 into 8 equal parts. How much is each part?

Number sentence: _____

5 Application problems.

(a) A primary school receives a donation of 720 books. The books are shared equally with 9 classes. How many books does each class receive?

Answer: _____

(b) The price of a large screen television is £1500. It is three times the price of a small screen television. How much does a small screen television cost?

£1500

Answer: _____

(c) In a hotel kitchen, 240 litres of cooking oil has been used from a large bottle. The amount of oil left in the bottle is 4 times the amount that has been used. How many litres of oil were in the bottle to start with?

Answer: _____

Challenge and extension question

6 Work out the number that each shape stands for.

▲ × ■ × ● = 540

▲ × ■ = 60

■ × ● = 180

(a) ▲ = [ ]   (b) ■ = [ ]   (c) ● = [ ]

9.11 Dividing a 2-digit number by a 1-digit number (1)

 Learning objective Divide 2-digit numbers by 1-digit numbers

 Basic questions

1 Calculate mentally.

(a) 9 ÷ 2 = ☐

(b) 25 ÷ 4 = ☐

(c) 27 ÷ 5 = ☐

(d) 38 ÷ 6 = ☐

(e) 19 ÷ 3 = ☐

(f) 36 ÷ 5 = ☐

(g) 39 ÷ 4 = ☐

(h) 47 ÷ 5 = ☐

2 What is the greatest number you can write in each box?

(a) 6 × ☐ < 32

(b) ☐ × 9 < 60

(c) ☐ × 8 < 53

(d) 8 × ☐ < 50

(e) 7 × ☐ < 62

(f) ☐ × 9 < 78

3 Five children share 73 sweets equally between them. How many sweets does each child have?

$73 \div 5 = ?$

(a) **Minna's solution:**

$5 \times 10 = \boxed{}$

$50 \div 5 = \boxed{}$

$73 - 50 = \boxed{}$

$23 \div 5 = \boxed{}$ r $\boxed{}$

$10 + 4 = \boxed{}$

$73 \div 5 = \boxed{}$ r $\boxed{}$

(b) **Asif's solution:**

$50 \div 5 = \boxed{}$

$23 \div 5 = \boxed{}$ r $\boxed{}$

$73 \div 5 = \boxed{}$ r $\boxed{}$

4 Fill in the boxes.

(a)

$64 \div 4 = \boxed{}$

$\boxed{40} \div \boxed{4} = \boxed{}$

$\boxed{24} \div \boxed{4} = \boxed{}$

(b)

$92 \div 4 = \boxed{}$

$\boxed{80} \div \boxed{4} = \boxed{}$

$\boxed{12} \div \boxed{4} = \boxed{}$

(c)

$84 \div 7 = \boxed{}$

$\boxed{} \div \boxed{} = \boxed{}$

$\boxed{} \div \boxed{} = \boxed{}$

(d)

$81 \div 7 = \boxed{}$ r $\boxed{}$

$\boxed{} \div \boxed{} = \boxed{}$

$\boxed{} \div \boxed{} = \boxed{}$ r $\boxed{}$

(e)

$86 \div 6 = \boxed{}$ r $\boxed{}$

$\boxed{} \div \boxed{} = \boxed{}$

$\boxed{} \div \boxed{} = \boxed{}$ r $\boxed{}$

5 Write a number sentence for each question.

(a) When 95 is divided by 7, what are the quotient and remainder?

Number sentence: _____

(b) A number times 5 is 75. What is the number?

Number sentence: _____

(c) What number divided by 5 is 125?

Number sentence: _____

6 Application problems.

(a) The weight of a calf is 6 kg. The weight of a cow is 90 kg. How many times the weight of the calf is that of the cow?

Answer: _____

(b) A toy factory plans to make 96 toys in 4 days. If it makes 25 toys each day, can the target be met?

Answer: _____

Challenge and extension question

7 Choose from these signs: +, −, ×, ÷ or ( ) to make these equations true. One has been done for you.

(a) 4 + 4 − 4 − 4 = 0   (b) 4   4   4   4 = 3

(c) 4   4   4   4 = 1   (d) 4   4   4   4 = 5

(e) 4   4   4   4 = 2   (f) 4   4   4   4 = 8

9.12 Dividing a 2-digit number by a 1-digit number (2)

Learning objective Divide 2-digit numbers by 1-digit numbers

Basic questions

1 Calculate mentally.

(a) $20 \div 6 =$ ☐   (b) $31 \div 5 =$ ☐   (c) $63 \div 8 =$ ☐

(d) $43 \div 7 =$ ☐   (e) $75 \div 9 =$ ☐   (f) $62 \div 7 =$ ☐

(g) $48 \div 9 =$ ☐   (h) $39 \div 4 =$ ☐

2 Use the column method to calculate. One has been done for you.

(a)
```
      1  5
  3 ) 4  5
      3
     ----
      1  5
      1  5
     ----
         0
```

(b)
```
  2 ) 7  8
```

(c)
```
  5 ) 6  5
```

(d)
```
  2 ) 4  6
```

(e)
```
  3 ) 9  3
```

(f)
```
  4 ) 4  8
```

3 Use the column method to calculate.

(a) $84 \div 7 =$ (b) $96 \div 3 =$ (c) $68 \div 2 =$ (d) $75 \div 3 =$

4 Write a number sentence for each question.

(a) What is 39 divided 3?

Number sentence: _____

(b) What is 78 divided by 6?

Number sentence: _____

5 Application problems.

(a) There are 35 white rabbits and 53 grey rabbits. If one rabbit hutch can keep 4 rabbits, how many hutches are needed to keep all the rabbits?

Answer: _____

(b) 54 children are dancing and they are divided into 2 rows. How many children are there in each row?

Answer: _____

 Challenge and extension questions

6 Using the numbers 4, 11, 2 and 3, complete the following division calculation.

☐ ÷ ☐ = ☐ r ☐

7 Lily's mother gave Lily 8 chocolates and gave the rest of the chocolates to Iram. The number of chocolates that Lily got was exactly half of the number Iram was given.

How many chocolates did Iram have?

Answer: _____

9.13 Dividing a 2-digit number by a 1-digit number (3)

Learning objective Divide 2-digit numbers by 1-digit numbers

Basic questions

1 What is the greatest number you can write in each box?

(a) ☐ × 7 < 45   (b) ☐ × 4 < 26   (c) 68 > 9 × ☐

(d) 3 × ☐ < 28   (e) 6 × ☐ < 35   (f) 47 > ☐ × 8

2 Work out these division calculations. Think carefully about where to write each quotient.

(a)

6) 3 0

_____

_____

(b)

2) 1 9

_____

_____

(c)

9) 3 8

_____

_____

(d)

3) 2 0

_____

_____

(e)

6) 3 2

_____

_____

(f)

4) 2 1

_____

_____

(g)

8) 2 8

_____

_____

(h)

7) 4 0

_____

_____

3 Use the column method to calculate.

(a) 27 ÷ 5 = ☐

(b) 40 ÷ 6 = ☐

(c) 58 ÷ 8 = ☐

(d) 66 ÷ 9 = ☐

(e) 54 ÷ 3 = ☐

(f) 65 ÷ 5 = ☐

(g) $64 \div 4 =$ ☐

(h) $72 \div 6 =$ ☐

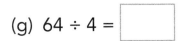

4 Application problems.

ART COMPETITION

(a) 27 pupils took part in a school art competition. They were divided into 3 groups. How many pupils were there in each group?

Answer: _____

(b) 88 balls are put into 5 baskets in equal numbers. How many balls are put into each basket? How many balls are left over?

Answer: _____

(c) Each coat needs 5 buttons. How many coats can 78 buttons be sewn on to? How many buttons are left over?

Answer: _____

Challenge and extension question

5 A teacher bought some special pencils as prizes for 6 winners of a mental maths competition. She gave each winner 1 pencil, and then continued to give them a pencil each until she no longer had enough pencils to give each winner. The number of pencils left over is the same as the number of pencils each winner has in total.

How many pencils did the teacher buy? Give as many answers as you can find.

9.14 Dividing a 2-digit number by a 1-digit number (4)

 Learning objective Divide 2-digit numbers by 1-digit numbers

Basic questions

1 Calculate mentally.

(a) 30 ÷ 3 = ☐

(b) 40 ÷ 2 = ☐

(c) 80 ÷ 4 = ☐

(d) 500 ÷ 5 = ☐

(e) 39 ÷ 3 = ☐

(f) 48 ÷ 2 = ☐

(g) 84 ÷ 4 = ☐

(h) 6000 ÷ 3 = ☐

2 Work out these calculations. (Note the differences in the results.)

(a) 96 ÷ 3 = ☐

(b) 92 ÷ 3 = ☐

(c) 96 ÷ 6 = ☐

(d) 92 ÷ 6 = ☐

3 Use the column method to calculate.

(a) 83 ÷ 4 = ☐

(b) 81 ÷ 2 = ☐

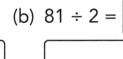

(c) 65 ÷ 6 = ☐

(d) 60 ÷ 3 = ☐

(e) 96 ÷ 9 = ☐

(f) 53 ÷ 5 = ☐

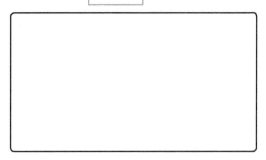

(g) 75 ÷ 7 = ☐

(h) 87 ÷ 8 = ☐

4 Complete the tables.

(a)

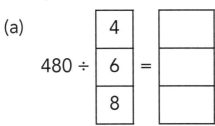

$$480 \div \begin{array}{|c|} \hline 4 \\ \hline 6 \\ \hline 8 \\ \hline \end{array} = \begin{array}{|c|} \hline \\ \hline \\ \hline \\ \hline \end{array}$$

(b)

$$3000 \div \begin{array}{|c|} \hline 3 \\ \hline 5 \\ \hline 6 \\ \hline \end{array} = \begin{array}{|c|} \hline \\ \hline \\ \hline \\ \hline \end{array}$$

5 Calculate mentally.

(a) $2 \times 200 = \boxed{}$ (b) $300 \times 3 = \boxed{}$ (c) $4 \times 500 = \boxed{}$

(d) $400 \div 2 = \boxed{}$ (e) $900 \div 3 = \boxed{}$ (f) $2000 \div 4 = \boxed{}$

6 Write a number sentence for each question.

(a) What is 82 divided by 2?

Number sentence: _____

(b) What is 900 divided by 3?

Number sentence: _____

(c) 640 is divided equally into 2 parts. How much is each part?

Number sentence: _____

(d) How many 4s are there in 320?

Number sentence: _____

7 A patch of grass the size of a classroom can produce enough oxygen daily for 3 people. How many of the patches of grass are needed to produce enough oxygen daily for 210 people?

Number sentence: _____

Challenge and extension question

8 A monkey is playing a game with 12 little monkeys to help them learn maths.

The rules are as follows.

(1) All the 12 little monkeys line up.

(2) The little monkeys in the 1st, 3rd, 5th, 7th, 9th and 11th places are given 1 banana each and then asked to leave. The other monkeys remain in the line.

(3) The little monkeys in the 1st, 3rd and 5th places in the new line are given 2 bananas each and then asked to leave. The other monkeys, again, remain in the line.

(4) The little monkeys in the 1st and 3rd places in the line are given 3 bananas each and then asked to leave the line.

(5) Finally, the remaining little monkey is given the top prize, which is 5 bananas.

If a monkey can choose any place in the line at the beginning, what place do you think it should choose in order to receive the top prize?

Answer: _____

9.15 Dividing a 2-digit number by a 1-digit number (5)

Learning objective Divide 2-digit numbers by 1-digit numbers

Basic questions

1 Calculate mentally.

(a) 30 × 2 = [ ]

(b) 120 × 3 = [ ]

(c) 9000 ÷ 3 = [ ]

(d) 2000 × 4 = [ ]

(e) 46 ÷ 2 = [ ]

(f) 300 × 5 = [ ]

(g) 60 ÷ 9 = [ ]

(h) 1000 × 7 = [ ]

(i) 21 × 3 = [ ]

(j) 840 ÷ 4 = [ ]

(k) 240 ÷ 6 = [ ]

(l) 2000 ÷ 5 = [ ]

2 Use the column method to calculate. Check your answers.

(a) 58 ÷ 4 = [ ]

(b) 61 ÷ 2 = [ ]

(c) 92 ÷ 5 = [ ]

(d) 37 ÷ 3 = ☐

(e) 97 ÷ 6 = ☐

(f) 74 ÷ 7 = ☐

3 Application problems.

(a) There are 96 sheep and 3 horses on a farm. How many times more sheep are there than horses?

Answer: _____

(b) A book has 63 pages. Mo is reading 7 pages a day. In how many days' time will he finish the book?

Answer: _____

(c) Maya helps her brother to ice some cupcakes. She ices 8 cupcakes every minute. How many minutes does it take her to ice 96 cupcakes?

Answer: _____

(d) 39 girls and 45 boys from Year 3 went to work in an orchard. They were divided into groups of 4. How many groups were they divided into?

Answer: _____

(e) Joe put 45 sweets into 15 sweet jars. He put 5 sweets in each jar. Did he have enough sweet jars? If your answer is yes, how many sweet jars were left over?

Answer: _____

Challenge and extension questions

4 Use the four numbers to write division sentences with remainders. Write as many as you can.

 3   **4**   **7**   **31**

5 Fill in the boxes.

(a) ☐ ÷ 7 = 5 r 4   (b) ☐ ÷ ☐ = 8 r 7

9.16 Dividing a 3-digit number by a 1-digit number (1)

 Learning objective Divide 3-digit numbers by 1-digit numbers

 Basic questions

1 Calculate mentally.

(a) $160 \div 8 =$ ☐

(b) $300 \div 6 =$ ☐

(c) $200 \div 5 =$ ☐

(d) $1600 \div 8 =$ ☐

(e) $3000 \div 6 =$ ☐

(f) $2000 \div 5 =$ ☐

(g) $220 \div 2 =$ ☐

(h) $840 \div 4 =$ ☐

2 597 books are shared equally among 4 classes. How many books does each class receive? How many books are left over?

$$597 \div 4 = \boxed{} \ r \ \boxed{}$$

(a) **Method 1:**

Since: $400 \div 4 =$ ☐

$160 \div 4 =$ ☐

$37 \div 4 =$ ☐

We have: $597 \div 4 =$ ☐

(b) **Method 2:**

Use the column method

☐

Each class receives ☐ books and ☐ book(s) is (are) left over.

(Check to see if the answer is correct.)

3 Complete these calculations.

(a)

637 ÷ 3 = ☐

600	÷	3	=	☐
30	÷	3	=	☐
7	÷	3	=	☐

(b)

665 ÷ 5 = ☐

500	÷	5	=	☐
150	÷	5	=	☐
15	÷	5	=	☐

(c)

738 ÷ 6 = ☐

☐	÷	☐	=	☐
☐	÷	☐	=	☐
☐	÷	☐	=	☐

4 Use the column method to calculate.

(a) 4) 5 3 6

(b) 7) 8 5 6

(c) 8) 9 3 4

5 Application problems.

    (a) A group of pupils picked 266 apples from the apple trees in the school's garden. 62 apples were saved for visitors. The rest were shared equally among the 6 classes in Year 1. How many apples did each class get?

Answer: _____

    (b) A school bought 4 boxes of books. Each box contains 100 books. How many books did the school buy in total? If the books are given to 5 year groups (from Year 1 to Year 5) equally, how many books will each year group receive?

Answer: _____

Challenge and extension question

6 Find the value of ▲ and fill in the boxes.

$$▲ + ▲ = \boxed{}$$

$$▲ - ▲ = \boxed{}$$

$$▲ × ▲ = \boxed{}$$

$$+ \quad ▲ ÷ ▲ = \boxed{}$$

$$\overline{}$$

$$100 \qquad ▲ = \boxed{}$$

9.17 Dividing a 3-digit number by a 1-digit number (2)

 Learning objective Divide 3-digit numbers by 1-digit numbers

 Basic questions

1 Calculate mentally.

(a) 80 ÷ 4 =

(b) 600 ÷ 6 =

(c) 510 − 480 =

(d) 39 ÷ 3 =

(e) 900 ÷ 3 =

(f) 60 − 37 =

(g) 23 × 4 =

(h) 25 × 6 =

2 What is the greatest number you can write in each box?

(a) 6 × ☐ < 38

(b) 5 × ☐ < 32

(c) 8 × ☐ < 85

(d) 4 × ☐ < 25

(e) 7 × ☐ < 60

(f) 3 × ☐ < 17

3 Try this on your own.

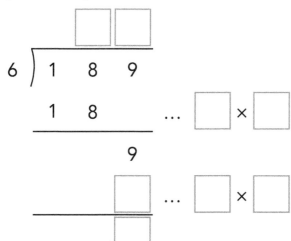

4 First decide how many digits the quotient has, and then calculate.

(a) $3 \overline{) 6 \quad 2 \quad 4}$

(b) $8 \overline{) 9 \quad 3 \quad 6}$

(c) $2 \overline{) 3 \quad 2 \quad 1}$

(d) $4 \overline{) 5 \quad 6 \quad 3 \quad 6}$

5 Complete these calculations and then check the answers.

(a) $656 \div 6 =$

(b) $736 \div 9 =$

(c) $496 \div 7 =$

6 Application problems.

(a) A dragonfly eats 4200 mosquitoes in a week. How many mosquitoes does it eat each day?

Answer: _____

(b) (i) 480 toy cars were divided equally into 5 large boxes. How many toy cars were placed into each large box?

Answer: _____

(ii) If the toy cars in each large box were put into 8 small boxes equally, how many toy cars were put into each small box?

Answer: _____

Challenge and extension questions

7 Think carefully. Which group do the following numbers belong to? Write the numbers in the correct circles.

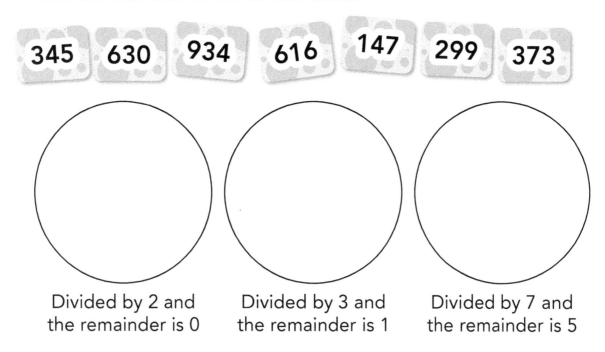

345   630   934   616   147   299   373

Divided by 2 and the remainder is 0

Divided by 3 and the remainder is 1

Divided by 7 and the remainder is 5

8 Omid was picking apples in an orchard. He picked 304 apples in 4 hours. Jenna picked 36 apples in half an hour. Who did the apple-picking faster? How many ways can you think of to help you to compare? Try to show two different ways.

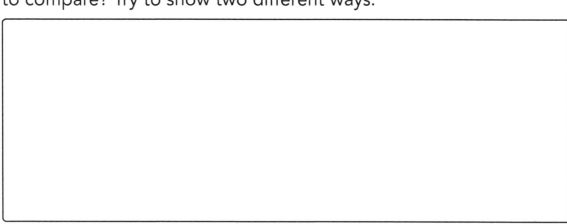

9.18 Dividing a 3-digit number by a 1-digit number (3)

 Learning objective Divide 3-digit numbers by 1-digit numbers

 Basic questions

1 Look at the number sentences in the first column and then complete the number sentences in the second column.

(a) 128 × 7 = 896   896 ÷ 7 = ☐

(b) 7254 ÷ 9 = 806   806 × 9 = ☐

2 Correct each calculation if it contains any mistakes.

(a) 354 ÷ 6 = 59   (b) 663 ÷ 7 = 99   (c) 428 ÷ 4 = 107

3 First, estimate how many digits the quotient will have and then use the column method to calculate. Check the answers to the questions marked with .

(a) 380 ÷ 3 =

(b) 843 ÷ 6 =

(c) 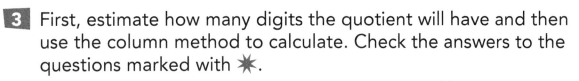 709 ÷ 7 =

(d)  450 ÷ 6 =

(e) 750 ÷ 5 =

(f) 919 ÷ 9 =

4 Fill in the boxes.

(a) ☐ × 7 = 266

(b) 9 × ☐ = 486

(c) ☐ ÷ 6 = 350

(d) 384 ÷ ☐ = 8

5 True or false? (Put a ✓ for true and a ✗ for false in each box.)

(a) 0 divided by any non-zero number is 0. ☐

(b) For division, ☐ 0 ☐ ÷ 4, only when the digit in the hundreds place of the dividend is 4, can the digit in the tens place of the quotient be 0. . . . . ☐

(c) If the divisor is a 1-digit number and the dividend is a 3-digit number, with 0 in its tens place, then the number in the tens place of the quotient must be 0. . . . . □

(d) If ● ÷ ▲ = 101 r 8, then ▲ = 9, ● = 909. . . . . □

6 A school has bought 504 pots of flowers and plans to arrange them in rows. Each row must have the same number of pots. Can you design a few different ways to arrange them? Fill in the boxes and then write the number sentences.

(a) Plan 1: Put [ ] pots for each row for [ ] rows.

Number sentence: _____

(b) Plan 2: Put [ ] pots for each row for [ ] rows.

Number sentence: _____

(c) Plan 3: Put [ ] pots for each row for [ ] rows.

Number sentence: _____

Challenge and extension question

7 Two school buildings are 80 metres apart. If one pot of flowers is placed every 2 metres, how many pots of flowers are needed to cover the distance from one building to the other?

Answer: _____

9.19 Application of division

Learning objective Use division to solve practical problems

Basic questions

1 Calculate mentally.

(a) 17 ÷ 3 = ☐

(b) 33 ÷ 7 = ☐

(c) 43 ÷ 8 = ☐

(d) 29 ÷ 9 = ☐

(e) 52 ÷ 6 = ☐

(f) 75 ÷ 10 = ☐

(g) 25 ÷ 4 = ☐

(h) 42 ÷ 5 = ☐

2 Complete the sentences.

(a) When 618 is divided by 5, the quotient is ☐ and the remainder is ☐ .

(b) To see if 251 ÷ 3 = 83 r 2 is correct, you can use

☐ × ☐ + ☐ = ☐ to check.

(c) There are ☐ zeros at the end of the product of 800 × 5.

(d) In a division sentence, if both the dividend and divisor are the same, then the quotient is ☐ . If the dividend and the quotient are the same, then the divisor is ☐ . If the quotient is to be 0, the dividend is ☐ .

3 Application problems.

(a) A group of pupils plan to put 140 kg of marbles into boxes. Each box can hold 6 kg. At least how many boxes are needed?

Answer: _____

(b) Two Year 3 classes go rowing. Each boat can seat 7 children. There are 29 pupils in Class One and 32 pupils in Class Two. How many boats are needed for each class if they go separately? If the classes share the boats, how many boats are needed?

Answer: _____

(c) Theo has £50 to buy some pencil boxes. Each pencil box costs £8. How many pencil boxes can he buy?

Answer: _____

(d) Ella has 147 sheets of paper. 8 sheets of paper are needed to make one exercise book.

(i) How many exercise books can she make with the 147 sheets of paper?

Answer: _____

(ii) With the remaining sheets of paper, how many more sheets are needed to make another exercise book?

Answer: _____

4 A company uses 3 kilograms of fresh fish to make 1 kilogram of dried fish. How many kilograms of dried fish can it make using 750 kilograms of fresh fish?

Answer: _____

5 Dylan asked Amina to work out a puzzle: 'A piece of wood is sawn into 7 pieces. It takes 5 minutes to saw each piece. How long does it take to saw all 7 pieces?'

Amina answered quickly: '5 minutes for 1 piece, of course – it takes 35 minutes to saw 7 pieces.'

'No, it takes 30 minutes!' said Dylan. Is he right? _____

9.20 Finding the total price

Learning objective Use division to solve practical problems

Basic questions

1 Calculate mentally.

(a) 21 × 3 = ☐

(b) 80 ÷ 5 = ☐

(c) 16 × 2 = ☐

(d) 40 × 3 − 40 = ☐

(e) 7200 ÷ 6 = ☐

(f) 20 × 4 = ☐

(g) 180 + 70 = ☐

(h) 49 ÷ 7 + 8 = ☐

(i) 13 × 7 = ☐

(j) 90 − 26 = ☐

(k) 270 ÷ 9 = ☐

(l) 48 ÷ 3 + 56 = ☐

2 Complete the table.

	Pencil case	Pen	Coloured pencil (in box)
Unit price (price per item)	£6		£9 per box
Quantity (number of items)	7	10	
Total price		£340	£900

3 Write × or ÷ in the ◯ and 'total price', 'unit price' or 'quantity' in the spaces.

(a) unit price ◯ _____ = total price

(b) total price ◯ _____ = quantity

(c) total price ◯ _____ = unit price

4 Meena bought 4 saris. Each sari costs £105.
How much did she spend in total?

Answer: _____

5 There are 24 apples in a box. The price of each box of apples is £30. What is the total price of 3 of the boxes? How many apples are there in 8 boxes?

Answer: _____

6 Write a number sentence for each question.

(a) What is 6 times 660?

Number sentence: _____

(b) How many times 6 is 660?

Number sentence: _____

(c) When the divisor is 8, the quotient is 402 and the remainder is 7, what is the dividend?

Number sentence: _____

7 A box can hold 8 basketballs. How many boxes are needed to hold 768 basketballs?

Answer: _____

 Challenge and extension questions

8 A school has planted 12 trees along one side of the playground, from one end to the other. One tree was planted every 6 metres. How long is the playground?

Answer: _____

9 It took Leila 5 minutes to walk from the 1st tree to the 6th tree along the main path in a school playground. Which tree did she reach after 15 minutes if she walked at the same pace?

Answer: _____

Chapter 9 test

1 Calculate mentally.

(a) $8 \times 30 = $ ☐

(b) $40 \times 40 = $ ☐

(c) $24 \times 5 - 100 = $ ☐

(d) $7 \times 8 + 45 = $ ☐

(e) $300 \times 6 = $ ☐

(f) $50 \div 10 = $ ☐

(g) $8 \times 2 \times 4 = $ ☐

(h) $80 \div 4 - 17 = $ ☐

(i) $540 + 5 = $ ☐

(j) $200 \div 4 = $ ☐

(k) $400 \div 8 = $ ☐

(l) $170 + 130 = $ ☐

(m) $5400 \div 6 = $ ☐

(n) $360 \div 4 = $ ☐

(o) $8 \times 2 + 12 = $ ☐

(p) $48 \div 8 + 9 = $ ☐

2 Use the column method to calculate. Check the answer to the question marked with ✳.

(a) $4 \times 276 = $

(b) $780 \div 6 = $

(c) ✳ $919 \div 9 = $

3 Work out these calculations, step by step.

(a) $182 \times 5 + 318$ (b) $456 \div 8 \times 4$ (c) $25 \times 4 + 175$

4 Write a number sentence for each question.

(a) 4 times a number is 520. What is the number?

Number sentence: _____

(b) What is the sum of 4 times 38 and the greatest 2-digit number?

Number sentence: _____

(c) The divisor is 7, the quotient is 192 and the remainder is 2. What is the dividend?

Number sentence: _____

(d) How many 3s are there in 69?

Number sentence: _____

5 Complete the sentences.

(a) The product of 48×6 is between ☐ and ☐, nearer to ☐.

(b) The product of the least 3-digit number and the greatest 2-digit number is ☐.

(c) $23 \times 6 = \boxed{} \times 6 + \boxed{} + 6 = \boxed{}$

$126 \times 4 = \boxed{} \times 4 + \boxed{} \times 4 + \boxed{} \times 4 = \boxed{}$.

(d) In $472 \div 8$, the greatest place of value of the quotient is in

the _____ place. It is a $\boxed{}$ -digit number.

(e) To make the quotient of ● $45 \div 5$ a 3-digit number, the smallest

possible number in the ● is $\boxed{}$. If the quotient is a 2-digit

number, the greatest possible number in the ● is $\boxed{}$.

(f) $829 \div \boxed{} = 5 \text{ r } 4$

(g) There are 20 red flowers. This is twice the number of yellow
flowers.

The total number of red flowers and yellow flowers is $\boxed{}$.

(h) In ▲ ÷ ● $= 16 \text{ r } 7$, the smallest possible number the divisor can

be is $\boxed{}$, which would make the dividend $\boxed{}$.

(i) There are $\boxed{}$ zeros at the end of the product 2400×5.

6 A school bought 8 computers. Each computer costs £650.
How much did the school pay in total for the computers?

Answer: _____

7 Mo and his 3 friends collected 634 empty bottles in a school environment campaign. If they put them into 8 bags equally, how many bottles were in each bag? How many empty bottles were left over?

Answer: _____

8 Jiro has 741 sheets of paper to make exercise books. 9 sheets of paper are needed to make one exercise book. How many exercise books can he make with the 741 sheets? With the remaining sheets of paper, how many more sheets are needed to make another exercise book?

Answer: _____

9 A fruit shop has 200 kg of watermelons, which is 5 times the amount of apples. How many more kilograms of watermelons does the shop have than that of apples?

Answer: _____

10 A group of children need to make 400 paper flowers for a nursery. They have made 260 flowers. The rest of the flowers must be made in two days. How many flowers do they need to make each day, on average, in the next two days?

Answer: _____

11 A toy factory has made 252 boxes of blocks in the past 9 days. Now it is making 35 boxes each day. How many more boxes of blocks does it make, on average, each day now than before?

Answer: _____

12 The weight of Alvin's uncle is 4 times that of Alvin. Alvin's weight is 24 kg. What is the weight of Alvin's uncle? How many more kilograms heavier is he than Alvin?

Answer: _____

13 In the number sentence, each ★ stands for the same digit. What digit does ★ stand for in order to make the number sentence correct?

$$1★ + ★1 + ★ = 11★$$

★ = ☐

Chapter 10 Let's practise geometry

10.1 Angles

Learning objective Identify and explore angles

Basic questions

1 How many angles are there in each shape? Write your answers below the shapes.

(a)   (b)   (c)   (d)

☐ angles   ☐ angles   ☐ angles   ☐ angles

2 True or false? (Put a ✓ for true and a ✗ for false in each box.)

(a) Every geometric shape must have at least one angle. ...... ☐

(b) The wider the two sides of an angle are
open, the greater the angle. .... ☐

(c) All right angles are equal. .... ☐

(d) Two right angles make a half turn. .... ☐

3 Multiple choice questions.

(a) In the angles below, the greatest angle is ☐ .

A.   B.   C.   D.

(b) When the hour hand and the minute hand on a clock face form a right angle, it could be ☐ .

   A. 12 o'clock   B. half past 3   C. 6 o'clock   D. 9 o'clock

(c) ☐ right angles make a complete turn.

   A. One   B. Two   C. Three   D. Four

4 How many angles are there in each shape? How many are right angles? Write your answers below the shapes.

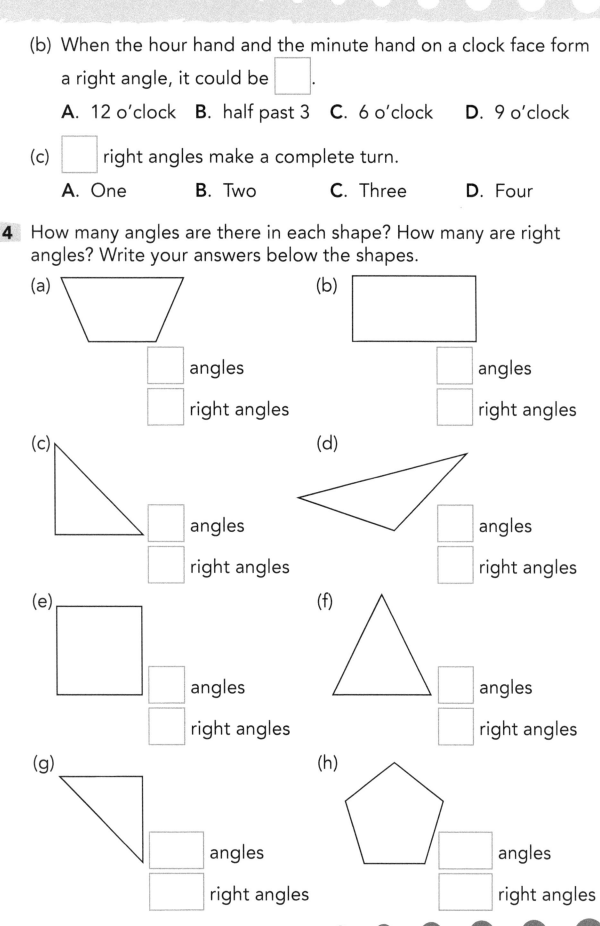

(a)
☐ angles
☐ right angles

(b)
☐ angles
☐ right angles

(c)
☐ angles
☐ right angles

(d)
☐ angles
☐ right angles

(e)
☐ angles
☐ right angles

(f)
☐ angles
☐ right angles

(g)
☐ angles
☐ right angles

(h)
☐ angles
☐ right angles

5 In the grid, draw the three angles that are described below.

(a) A right angle

(b) An angle that is less than a right angle (also called 'an acute angle')

(c) An angle that is greater than a right angle (also called 'an obtuse angle')

Challenge and extension questions

6 Use a ruler to draw a line on the shape so that it has 3 more right angles.

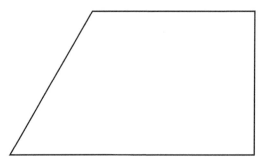

7 A square has 4 angles. If one of its 4 corners is cut off, how many angles will there be in the remaining part? (You may draw squares on paper and then cut them out to help you find the answer.)

10.2 Identifying different types of line (1)

Learning objective Identify horizontal and vertical lines

 Basic questions

1 Draw lines to match the names and pictures to the descriptions.

| Horizontal lines | | Lines that run from top to bottom | | │ │ │ |

| Vertical lines | | Lines that run from left to right | | ─── ─── ─── |

2 Look at the shapes. Write the numbers of the vertical lines and the horizontal lines in the spaces.

(a)   (b)

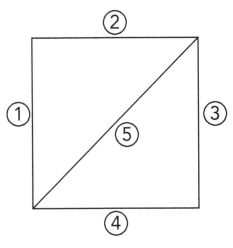

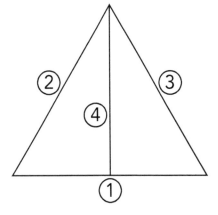

Vertical line(s): _____   Vertical line(s): _____

Horizontal line(s): _____   Horizontal line(s): _____

3 The picture below shows one side of a house. Find all the vertical lines and number them. Write the numbers in the box below. Then do the same with all the horizontal lines. One has been done for you.

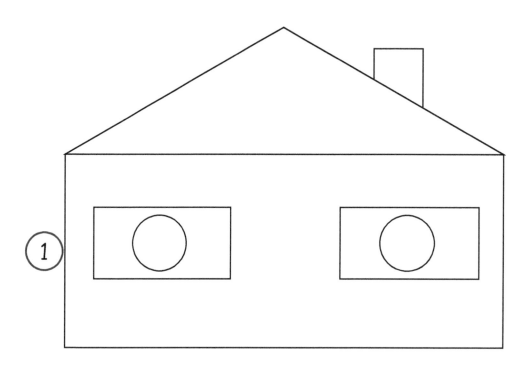

Vertical line numbers: 1,_____

Horizontal line numbers: _____

4 True or false? (Put a ✓ for true and a ✗ for false in each box.)

(a) After a quarter turn, a horizontal line is still a horizontal line. ⬚

(b) After a half turn, a vertical line will be a horizontal line. ⬚

(c) After a three quarter turn, a horizontal line will be a vertical line. ⬚

(d) After a complete turn, a vertical line is still a vertical line. ⬚

Challenge and extension question

5 Write 2–4 examples of vertical and horizontal lines from your everyday life. (Hint: Look around your home, your classroom or the environment for ideas.)

_____

_____

_____

_____

_____

_____

_____

_____

_____

_____

10.3 Identifying different types of line (2)

 Learning objective Identify perpendicular and parallel lines

 Basic questions

1 Draw lines to match the names and pictures to the descriptions.

Perpendicular lines

| Lines that will never meet | |

Parallel lines

| Lines that meet at a right angle | |

2 Identify the parallel and perpendicular lines in each shape. Use numbers to label the lines and write them in the spaces below. Write 'none' if there aren't any. One has been done for you.

(a)

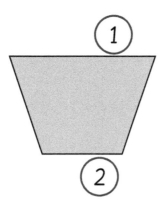

(b)

Parallel: 1, 2 _____

Perpendicular: _____

Parallel: _____

Perpendicular: _____

(c)

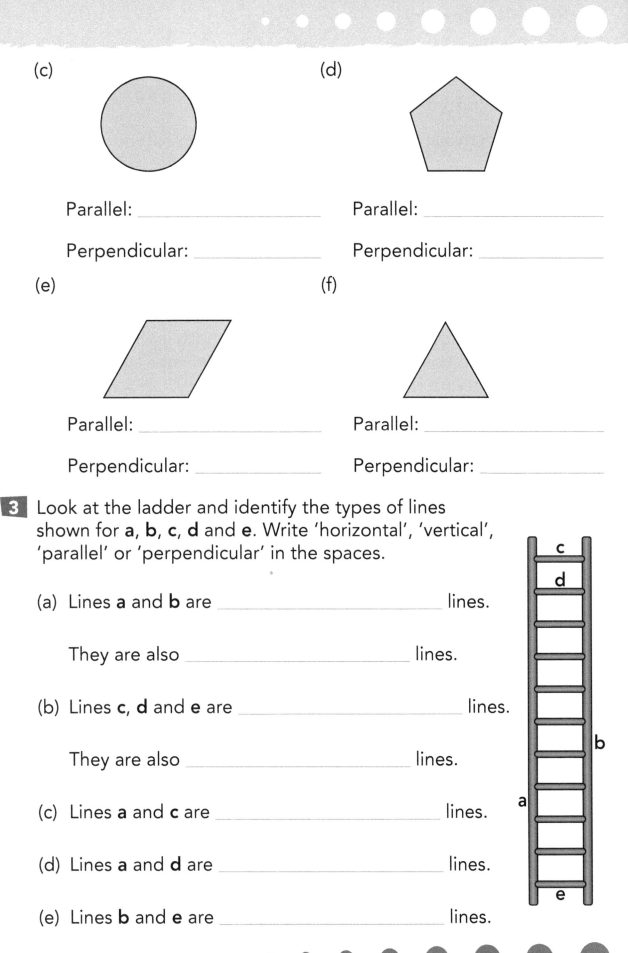

Parallel: _____

Perpendicular: _____

(d)

Parallel: _____

Perpendicular: _____

(e)

Parallel: _____

Perpendicular: _____

(f)

Parallel: _____

Perpendicular: _____

3 Look at the ladder and identify the types of lines shown for **a**, **b**, **c**, **d** and **e**. Write 'horizontal', 'vertical', 'parallel' or 'perpendicular' in the spaces.

(a) Lines **a** and **b** are _____ lines.

They are also _____ lines.

(b) Lines **c**, **d** and **e** are _____ lines.

They are also _____ lines.

(c) Lines **a** and **c** are _____ lines.

(d) Lines **a** and **d** are _____ lines.

(e) Lines **b** and **e** are _____ lines.

4 True or false? (Put a ✓ for true and a ✗ for false in each box.)

(a) Two horizontal lines are parallel. . . . . □

(b) Two vertical lines are perpendicular to each other. . . . . □

(c) Two perpendicular lines will meet at a point. . . . . □

(d) Two lines are either parallel or perpendicular. . . . . □

Challenge and extension question

5 Write 2–4 examples of parallel and perpendicular lines from your everyday life. (Hint: Look around your home, your classroom or the environment for ideas.)

_____

_____

_____

_____

_____

_____

_____

_____

_____

10.4 Drawing 2-D shapes and making 3-D shapes

 Learning objective Explore properties of 2-D and 3-D shapes

 **Basic questions**

1 Complete the table. The first row has been done for you.

	Is it a 2-D or 3-D shape?	What is the name of the shape?	If it is a 2-D shape, is it a symmetrical figure?
(triangle)	2-D	triangle	no
(cylinder)			
(square)			
(cuboid)			
(hexagon)			
(cone/pyramid)			

2 Use a straight edge to draw these 2-D shapes.

   (a) a rectangle   (b) a pentagon

   (c) an octagon

3 Use the 1 cm square dot grids to draw the shapes described.

(a) A triangle with one angle a right angle and the lengths of two of its sides 2 cm and 3 cm

(b) A hexagon with two sides parallel and the lengths of the two parallel sides 1 cm and 5 cm

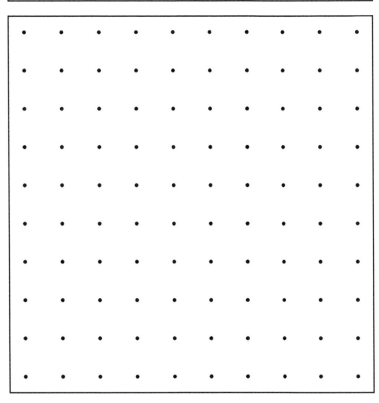

4 Use square dot grid paper to draw a 2-D shape consisting of six squares as shown below (drawing not to scale). Then cut it out carefully and make a 3-D shape. What shape do you get?

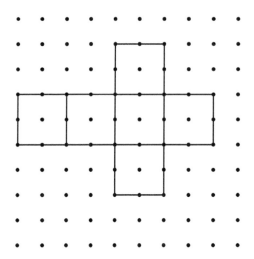

Answer: _____

 Challenge and extension question

5 Draw the 2-D shape below on a piece of paper (not necessarily to the same scale), cut it out carefully and make a 3-D shape. What shape do you get?

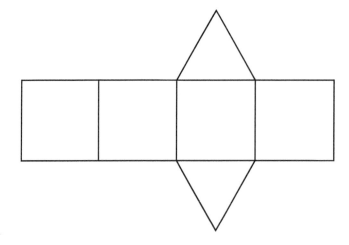

Answer: _____

10.5 Length: metre, centimetre and millimetre

 Learning objective Measure and compare lengths using metres, centimetres and millimetres

 Basic questions

1 Fill in the boxes.

(a) 1 m = ☐ cm

(b) 1 cm = ☐ mm

(c) 1 m = ☐ mm

(d) $\frac{1}{2}$ m = ☐ cm

(e) 15 cm = ☐ mm

(f) 700 cm = ☐ m

(g) 1 m and 30 cm = ☐ cm

(h) 2 m and 26 cm = ☐ cm

(i) $\frac{3}{10}$ cm = ☐ mm

(j) 900 mm = ☐ cm

2 How long is each object? Write the answers in the boxes.

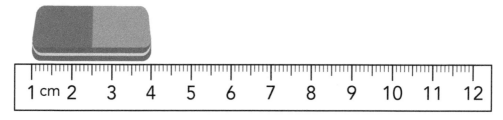

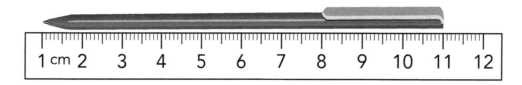

(a) The rubber is [ ] cm long.

(b) The pen is [ ] mm long.

(c) The pencil is [ ] cm and [ ] mm long.

3 Write a suitable unit in each box.

(a) A dining table is about 80 [ ] high.

(b) A nail is about 50 [ ] long.

(c) A giraffe is about 5 [ ] tall.

(d) The length of a bus is about 10 [ ].

(e) A UK passport photo size is 45 [ ] high by 35 [ ] wide.

4 Write >, < or = in each ◯.

(a) 810 cm ◯ 8 m

(b) 7 m ◯ 75 cm

(c) 1 m ◯ 100 mm

(d) 5 cm ◯ 500 mm

(e) 6 m and 57 cm $\bigcirc$ 675 cm   (f) 360 cm $\bigcirc$ 3 m and 600 mm

(g) 28 m + 500 cm $\bigcirc$ 33 m   (h) 1 m + 60 mm $\bigcirc$ 106 cm

5 Write the following lengths in order, from the shortest to the longest. Write < to show the order.

(a)

| 340 cm | 12 m | 11 m 98 cm | 100 cm | 300 mm |

Answer: _____

(b)

| 50 m | 490 mm | 4 m and 90 cm | 600 cm |

Answer: _____

6 Application problems.

(a) A ribbon is 805 cm long. It is 1 m and 5 mm longer than another ribbon. What is the length of the other ribbon?

Answer: _____

(b) A giraffe is 4 m and 50 cm tall. It is 5 times as tall as an antelope. What is the height of the antelope?

Answer: _____

Challenge and extension question

7 (a) Jaya is 1 m and 65 cm tall. She is 30 cm taller than Mee. Theo is 10 cm taller than Mee. What is the height of Theo?

Answer: _____

(b) A 5-metre-long measuring pole was put into a river. The part of the pole above the water was 200 cm and the part of the pole in the mud at the bottom was 90 mm. How deep was the water in the river?

Answer: _____

10.6 Perimeters of simple 2-D shapes (1)

Learning objective Calculate and measure the perimeters of simple 2-D shapes

Basic questions

1 Use a coloured pen to trace the outline of each shape below.

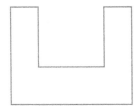  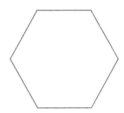

2 Find the perimeter of each shape on the 1 cm square grid paper.

Shape 1   Shape 2

Shape 3   Shape 4

(a) Shape 1: The perimeter

is _____.

(b) Shape 2: The perimeter

is _____.

(c) Shape 3: The perimeter

is _____.

(d) Shape 4: The perimeter

is _____.

Let's practise geometry

3 Find the perimeter of each shape. The drawings are not to scale.

(a)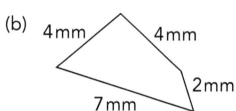
5 cm   5 cm

7 cm

The perimeter is _____.

(b)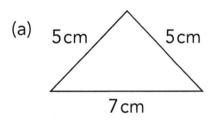
4 mm   4 mm

2 mm

7 mm

The perimeter is _____.

(c)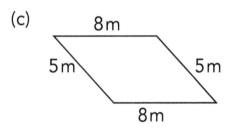
8 m

5 m   5 m

8 m

The perimeter is _____.

4 Every morning Darshan runs 6 laps on a path around a pond near his house. The pond is a pentagon shape and each side is 30 metres long. What distance does he run every morning?

Answer: _____

5 Look at the diagram. Five identical squares each with sides 1 cm in length are pieced together so that each square has at least one side touching the side of another square.

How many different arrangements are possible, not including the shape shown?

Answer: _____

Out of all the possible arrangements of the five squares, what is the shortest perimeter?

Answer: _____

10.7 Perimeters of simple 2-D shapes (2)

 Learning objective Calculate and measure the perimeters of simple 2-D shapes

 Basic questions

1 Find the perimeter of each shape. The length of one side of each small square is 1 cm.

(a)

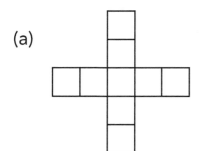

Perimeter: _____.

(b)

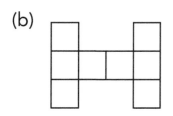

Perimeter: _____.

(c)

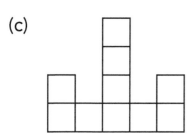

Perimeter: _____.

(d)

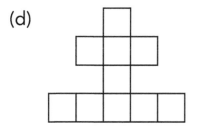

Perimeter: _____.

2 Calculate the perimeter of each of these shapes.
The drawings are not to scale.

(a)

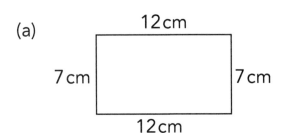

Perimeter: _____.

(b)

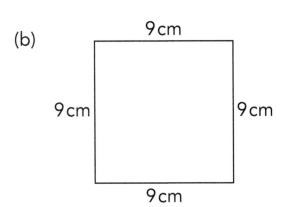

Perimeter: _____.

(c)

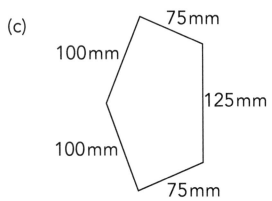

Perimeter: _____.

(d)

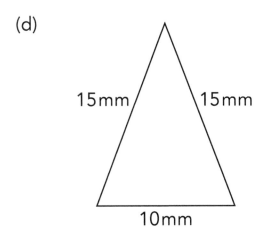

Perimeter: _____.

3 Measure each of the shapes below and find its perimeter in mm. Write the number sentence you use to find the answer.

(a)

Number sentence: _____

Answer: _____

(b)

Number sentence: _____

Answer: _____

(c)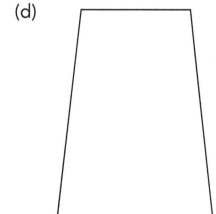

Number sentence: _____

Answer: _____

(d)

Number sentence: _____

Answer: _____

4 Look at the diagram. An ant is walking along the lines of an 8 cm square grid from Point A to Point B. At least how many centimetres must the ant walk to reach Point B?

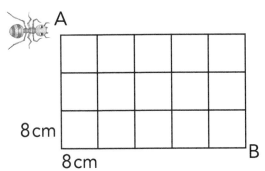

Answer: _____

Challenge and extension questions

5 The figure below shows a rectangle. A curve connecting A and B divides the rectangle into two shapes. Does Shape 1 have a longer perimeter? Circle 'yes' or 'no' and give your reason.

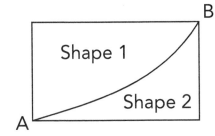

yes   no

Reason: _____

_____

_____

6 What is the perimeter in mm of each of the shapes below? Write the number sentence you use to find the answer.

(a)

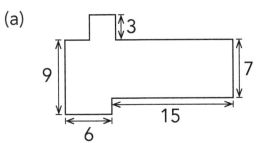

(b)

Number sentence:

_____

Answer: _____

Number sentence:

_____

Answer: _____

Chapter 10 test

1 Are all of the following angles? Write each number that shows an angle and a right angle on the answer lines below.

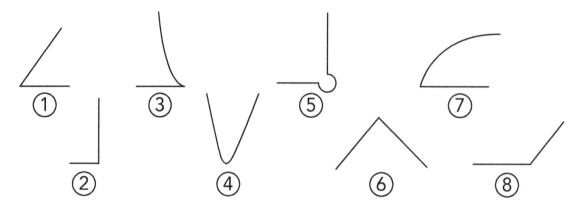

Angle(s): _____ Right angle(s): _____

2 True or false? (Put a ✓ for true and a ✗ for false in each box.)

(a) All right angles are equal. . . . . ☐

(b) Some geometric shapes do not have angles. . . . . ☐

(c) After a quarter turn, a vertical line will
     be a horizontal line. . . . . ☐

(d) After a complete turn, a horizontal line
     will be a vertical line. . . . . ☐

(e) Two perpendicular lines meet at a right angle. . . . . ☐

(f) If two lines are not parallel, they must be perpendicular. . . . . ☐

(g) The length all the way round a swimming pool is
     the perimeter of the pool. . . . . ☐

(h) If the lengths of the three sides of a triangle
     are 3 cm, 4 cm and 5 cm, its perimeter is 60 cm. . . . . ☐

3 Look at the picture then complete the sentences by writing 'horizontal', 'vertical', 'parallel' or 'perpendicular'.

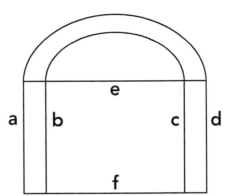

(a) Lines **a**, **b**, **c** and **d** are _____ lines.

They are also _____ lines.

(b) Lines **e** and **f** are _____ lines.

They are also _____ lines.

(c) Lines **a** and **f** are _____ lines.

(d) Lines **a** and **e** are _____ lines.

(e) Lines **b** and **f** are _____ lines.

4 Fill in the boxes.

(a) $8\,m =$ ☐ cm

(b) $1\,m =$ ☐ mm

(c) $900\,mm =$ ☐ cm

(d) $\frac{1}{2}\,cm =$ ☐ mm

(e) $100\,cm =$ ☐ mm

(f) $1000\,cm =$ ☐ m

(g) $\frac{1}{4}\,m =$ ☐ mm

(h) $5\,m$ and $66\,cm =$ ☐ cm

(i) $880\,cm =$ ☐ m and ☐ cm

(j) $\frac{9}{10}\,cm =$ ☐ mm

5 Multiple choice questions.

(a) The figure that shows a right angle is ☐.

A. B. C. D.

(b) Each side of a square is 4 cm long. Its perimeter is ☐.

A. 4 cm   B. 8 cm   C. 12 cm   D. 16 cm

(c) Nine identical squares, each with side length 1 cm, are put together to form the following shapes.

The shape with the shortest perimeter is ☐.

A. B. C. D.

(d) Look at the diagram and compare the perimeters of Shape A and Shape B.

The correct statement is ☐.

A. They are equal.

B. The perimeter of Shape A is greater than that of Shape B.

C. The perimeter of Shape A is shorter than the perimeter of Shape B.

D. Not sure.

Shape A

Shape B

6 Count the angles in the diagram.

There are ☐ angles.

7 Use the 1 cm square dot grids to draw the shapes described.

(a) A square with side length 3 cm and all the sides either horizontal or vertical

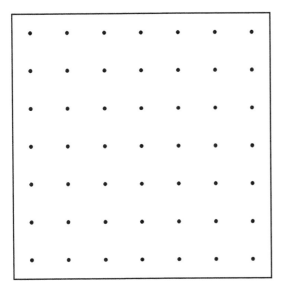

(b) A pentagon with the bottom side 3 cm long and horizontal, and no vertical sides

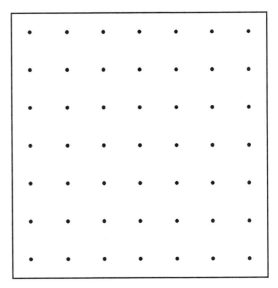

8 Look at the shaded figure on the 1 m square grid (drawing not to scale). What is the perimeter of the shaded figure?

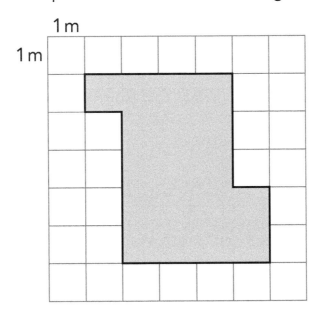

Answer: _____

9 Measure each shape and find its perimeter in mm. Write the number sentence you use to get your result. (unit: mm)

(a)

Number sentence: _____

Answer: _____

(b)

Number sentence: _____

Answer: _____

(c)

Number sentence: _____

Answer: _____

10 Keir accidently spilled ink over a piece of graph paper, as shown here. If the length of one side of each small square is 2 cm, how long is the perimeter of the piece of paper?

Answer: _____

11 Every day, Lala runs 3 complete laps around a hexagonal path near her home. Each side of the hexagonal path is 90 metres long. How many metres does she run every day?

Answer: _____

End of year test

1 Calculate mentally. (12%)

(a) 600 + 200 = ☐

(b) 150 − 90 = ☐

(c) 24 × 5 = ☐

(d) 54 ÷ 9 = ☐

(e) 248 + 352 = ☐

(f) 835 − 738 = ☐

(g) 270 ÷ 3 = ☐

(h) 770 ÷ 7 = ☐

(i) 6 × 9 + 7 = ☐

(j) 7 × 6 × 0 = ☐

(k) 7 × 5 × 2 = ☐

(l) 25 × 4 + 105 = ☐

2 Find the answers to these calculations. (6%)

(a) $\frac{1}{5} + \frac{3}{5} =$ ☐

(b) $\frac{5}{9} - \frac{2}{9} =$ ☐

(c) $1 - \frac{1}{6} =$ ☐

(d) $\frac{5}{11} + \frac{3}{11} =$ ☐

(e) $\frac{3}{8} - \frac{3}{8} =$ ☐

(f) $\frac{4}{7} + 0 =$ ☐

3 Multiple choice questions. (8%)

(a) The product of two sevens is ☐.

    A. 27   **B.** 14   **C.** 49   **D.** unknown

(b) ☐ of these show $\frac{1}{4}$ of the whole shaded.

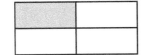 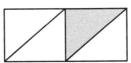  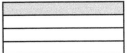

    A. One   **B.** Two   **C.** Three   **D.** All

(c) Kim started lunch at five to twelve and took 32 minutes to finish. She finished at ☐.

    A. 12:32   **B.** 12:27   **C.** 11:23   **D.** 12:37

(d) When both the divisor and the quotient are 9, the dividend is ☐.

    A. 9   **B.** 18   **C.** 27   **D.** 81

4 Write the answers. (16%)

(a) 1 m = ☐ cm = ☐ mm

(b) Write a suitable unit for each of these.

  (i) Ethan's weight is 30 _____.

  (ii) Milly did 15 skips in 5 _____.

  (iii) A textbook costs 7 _____.

  (iv) In a leap year, there are 29 _____ in February.

(c) After a 1-metre-long rope is folded in half twice, each part is _____ of the rope. It is _____ metres long.

(d) $\dfrac{2}{3} = \dfrac{4}{\Box} = \dfrac{\Box}{12}$

(e) Put these five numbers in order from the greatest to the least.

799   908   1000   857   988

☐ ☐ ☐ ☐ ☐

(f) Given ■ ÷ ● = 3 r 6, the smallest possible number that ● can be is ☐ , which would mean ■ must be ☐ .

(g) A leap year has ☐ days, which is ☐ weeks and ☐ days.

5 Use the column method to work out the answers. (18%)

(a) 187 + 552 = ☐

(b) 607 − 268 = ☐

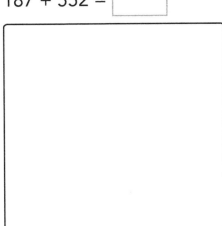

(c) 1000 − 591 = ☐

(d) 3 × 278 = ☐

(e) 124 × 8 = ☐

(f) 612 ÷ 9 = ☐

6 The diagram below consists of three squares. Answer the following questions. (6%)

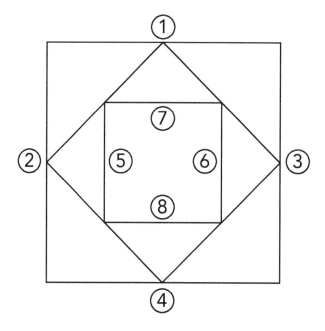

(a) How many right angles are there in total in the diagram?

(b) Write the numbers of all the vertical lines.

_____

(c) Write the numbers of all the horizontal lines.

_____

7 Find the perimeter of each shape. (8%)

   (a) Use metres as the unit.   (b) Use centimetres as the unit.

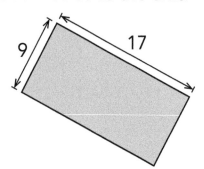

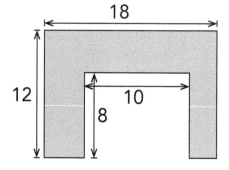

   Perimeter = _____   Perimeter = _____

8 Application problems. (26%. 4% each for questions (a) to (d); 10% for question (e))

(a) Aliya is as tall as her brother when she stands on a chair. Aliya is 125 cm tall. Her brother is 180 cm tall. What is the height of the chair?

Number sentence: _____

(b) A box of sweets is shared by 9 children equally. Each child gets 5 sweets and there are 6 sweets left over. How many sweets were there in the box to start with?

Number sentence: _____

(c) A road is 287 metres long. A gardener plants 8 trees equally spaced from one end of the road to the other. What is the distance between two neighbouring trees?

Number sentence: _____

(d) There are 152 boys in Year 2. The number of girls is 16 fewer than twice the number of boys. How many girls are there in Year 2?

Number sentence: _____

(e) The table shows the number of pupils in Year 3 classes.

Class	Class 1	Class 2	Class 3	Class 4
Number of pupils	28	26	30	32

(i) Make a bar chart based on the data shown in the table above.

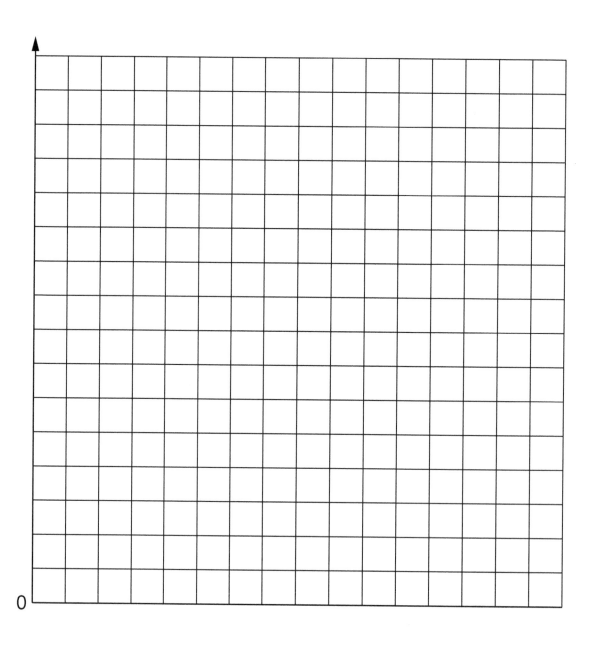

0

(ii) Based on the bar chart, Class ____ has the most pupils and

Class ____ has the fewest pupils.

The difference is ⬜ pupils.

(iii) There are ⬜ pupils in total in these four classes.

(iv) Write a question based on the bar chart and then answer it.

Question: _____

_____

_____

_____

Answer: _____

_____

_____

_____

Notes

Notes

Notes

Notes

Notes

Notes

Notes

Notes

Notes